T0248591

Pitman Research Notes in Mathematics Series

Submission of proposals for consideration

Suggestions for publication, in the form of outlines and representative samples, are invited by the Editorial Board for assessment. Intending authors should approach one of the main editors or another member of the Editorial Board, citing the relevant AMS subject classifications. Alternatively, outlines may be sent directly to the publisher's offices. Refereeing is by members of the board and other mathematical authorities in the topic concerned, throughout the world.

Preparation of accepted manuscripts

On acceptance of a proposal, the publisher will supply full instructions for the preparation of manuscripts in a form suitable for direct photo-lithographic reproduction. Specially printed grid sheets can be provided and a contribution is offered by the publisher towards the cost of typing. Word processor output, subject to the publisher's approval, is also acceptable.

Illustrations should be prepared by the authors, ready for direct reproduction without further improvement. The use of hand-drawn symbols should be avoided wherever possible, in order to maintain maximum clarity of the text.

The publisher will be pleased to give any guidance necessary during the preparation of a typescript, and will be happy to answer any queries.

Important note

In order to avoid later retyping, intending authors are strongly urged not to begin final preparation of a typescript before receiving the publisher's guidelines. In this way it is hoped to preserve the uniform appearance of the series.

Longman Scientific & Technical
Longman House
Burnt Mill
Harlow, Essex, CM20 2JE
UK
(Telephone (0279) 426721)

James H Bramble

Cornell University, USA

Multigrid methods

CRC Press
Taylor & Francis Group
Boca Raton London New York

CRC Press is an imprint of the
Taylor & Francis Group, an **informa** business

A CHAPMAN & HALL BOOK

Chapman and Hall/CRC
Taylor & Francis Group
6000 Broken Sound Parkway NW, Suite 300
Boca Raton, FL 33487-2742

© 1993 by Taylor & Francis Group, LLC
CRC Press is an imprint of Taylor & Francis Group, an Informa business

First issued in paperback 2019

No claim to original U.S. Government works

ISBN 13: 978-0-367-44971-1 (pbk)
ISBN 13: 978-0-582-23435-2 (hbk)

Visit the Taylor & Francis Web site at
http://www.taylorandfrancis.com

and the CRC Press Web site at
http://www.crcpress.com

Table of Contents

Preface

These notes are based on graduate courses given at Cornell University in 1985, 1988, 1989 and 1992. They are meant to introduce the reader to theoretical considerations involved in the study of multigrid methods, always keeping the practical applicability in mind. The aim is to present some of the most recent results in an abstract setting. The fundamental ingredients are separated from each other and are presented in the form of conditions. For instance, conditions required of the smoothing operator are kept separate from intrinsic properties (e.g. regularity properties) inherent in the problem to be treated. These conditions are listed in Appendix C as a glossary for convenient reference.

In these notes only symmetric and positive definite problems are considered, and no attempt is made to cover the vast literature on the subject of multilevel methods.

The following is an outline of the topics covered in these notes. Section 1 is a brief introduction to symmetric linear iterative methods. In Section 2 we consider a model problem and discuss a two level method. The third section contains many abstract results concerning multigrid algorithms in the case of nested spaces and inherited forms. Section 4 is devoted to a more general abstract setting. Here, nonnested spaces as well as forms which may vary from level to level are covered. General algorithms are formulated which include multiple smoothings which may vary in number from level to level, as well as multiple corrections.

The next section concerns the construction of two classes of commonly used smoothing operators and discusses the conditions under which they satisfy the conditions required of the abstract theory.

Sections 6–9 contain examples of specific applications of the theoretical results. These applications include second order elliptic problems with curved boundaries, mesh refinements and numerical quadrature. Also considered are first kind boundary integral operators. The final applications are to the biharmonic Dirichlet problem and include the Morley nonconforming method and the Ciarlet–Raviart mixed finite element method.

Section 10 contains a discussion of how the general theory is implemented to obtain practical algorithms in specific cases. This section also contains work estimates; a discussion of the idea of "full multigrid" is presented there.

There are three appendices. Appendix A contains a derivation of the important conjugate gradient method. Error estimates for its convergence as an iterative method are derived, and a form of the algorithm is given which includes preconditioning.

Appendix B is a brief introduction to the theory of interpolation spaces, and theorems which are important for our applications are proved there. The last appendix is a "Glossary of Conditions". This is included as a convenient reference for the statements of the earlier theorems and their applications.

Finally, a substantial bibliography is included.

I especially want to thank Drs. Joseph Pasciak and Mark Hanisch and Professor Vidar Thomée. Dr. Hanisch was largely responsible for Section 9 which contains the applications to biharmonic problems. Dr. Pasciak made many suggestions throughout the entire manuscript and was most helpful in the presentation of the implementation issues of Section 10. Professor Thomée's constructive criticism of much of the manuscript, which greatly influenced the presentation, is very much appreciated.

I also want to thank Mr. Yaoping Zhang, who is currently a research assistant, for proofreading the entire manuscript. Finally, the expert and incredibly adept typing of the TeX files by Mrs. Arletta Havlik has been invaluable.

1. Introduction to Iterative Methods

We shall consider the equation

$$(1.1) \qquad\qquad Ax = f$$

in a finite dimensional real inner product space M with inner product $(\cdot,\cdot)$ and dimension N. The operator A will be linear, symmetric with respect to $(\cdot,\cdot)$ and positive definite (SPD).

For such operators A we have the eigenvalues $\{\Lambda_i\}$ and corresponding eigenvectors $\{\varphi_i\}$. That is, $0 < \Lambda_1 \le \Lambda_2 \le \cdots \le \Lambda_N$ and

$$A\varphi_i = \Lambda_i \varphi_i, \quad i = 1,\ldots,N.$$

We may take φ_i so that

$$(\varphi_i, \varphi_j) = \delta_{ij}$$

where δ_{ij} is the Kronecker delta. The eigenvectors form an orthonormal basis for M, i.e., for any $u \in M$

$$(1.2) \qquad\qquad u = \sum_{i=1}^{N} (u, \varphi_i)\varphi_i.$$

We want to consider the following linear iterative methods. Let B be another operator on M which is SPD. Consider for x^0 arbitrary,

$$(1.3) \qquad\qquad x^{n+1} = x^n - B(Ax^n - f).$$

This iteration is called linear, since the error operator $I - BA$ is linear. Such a scheme is automatically consistent in the sense that x is a fixed point.

We consider first the simple case $B = \tau I$, with τ a scalar and examine the convergence properties of (1.3) for this example.

To do this set $e^n = x - x^n$. Then

(1.4)
$$e^n = e^{n-1} - \tau A e^{n-1} = (I - \tau A)e^{n-1}$$
$$= (I - \tau A)^n e^0.$$

Now, since the eigenvectors of A form an orthonormal basis for M, we have

$$e^0 = \sum_{i=1}^{N} (e^0, \varphi_i)\varphi_i.$$

Hence

$$(I - \tau A)^n e^0 = \sum_{i=1}^{N} (e^0, \varphi_i)(1 - \tau \Lambda_i)^n \varphi_i$$

and

(1.5)
$$\|(I - \tau A)^n e^0\|^2 = \sum_{i=1}^{N} (1 - \tau \Lambda_i)^{2n} (e^0, \varphi_i)^2$$
$$\leq \sup_{\Lambda_i \in \sigma(A)} |1 - \tau \Lambda_i|^{2n} \|e^0\|^2,$$

where $\|\cdot\| = (\cdot, \cdot)^{1/2}$ and $\sigma(A)$ denotes the spectrum of A. Thus the operator norm of $I - \tau A$ satisfies

$$\|I - \tau A\| \leq \sup_{\Lambda \in [\Lambda_1, \Lambda_N]} |1 - \tau \Lambda|.$$

Clearly $\tau = \frac{2}{\Lambda_1 + \Lambda_N}$ is the best choice in the sense that it minimizes $\|I - \tau A\|$.

The simple choice $\tau = \Lambda_N^{-1}$ tells us what is happening. Looking at $e^n = \sum_{i=1}^{N} (e^0, \varphi_i)(1 - \Lambda_N^{-1} \Lambda_i)^n \varphi_i$, we see that the components of the error, for $\Lambda_i \approx \Lambda_N$, are reduced well by this process. However, if $\Lambda_1 << \Lambda_N$ then $1 - \Lambda_N^{-1} \Lambda_1$ is near 1 and little happens to the low frequencies.

Define $K(A)$, the "condition number of A", to be $K(A) = \Lambda_N / \Lambda_1$. Now

$$\Lambda_1 \leq \frac{(A\varphi, \varphi)}{(\varphi, \varphi)} \leq \Lambda_N;$$

i.e., the largest and smallest eigenvalues are extremes of the Rayleigh quotient. Hence if we find numbers c_0 and c_1 such that

$$c_0(\varphi, \varphi) \leq (A\varphi, \varphi) \leq c_1(\varphi, \varphi)$$

then

$$c_0 \leq \Lambda_1, \quad \Lambda_N \leq c_1,$$

so that

$$K(A) \leq c_1/c_0$$

is an upper bound.

With the best choice of $\tau = \frac{2}{\Lambda_1 + \Lambda_N}$

$$\rho = \sup_{\Lambda \in [\Lambda_1, \Lambda_N]} |1 - \tau \Lambda| = 1 - \frac{2\Lambda_1}{\Lambda_1 + \Lambda_N} = \frac{\Lambda_N - \Lambda_1}{\Lambda_N + \Lambda_1} = \frac{K - 1}{K + 1}.$$

Note that this is not much of an improvement.

We next want to consider a more general iteration than that of the previous example. As in (1.3), define for x^0 arbitrary,

$$x^{n+1} = x^n - B(Ax^n - f)$$

where B is SPD. Define a new inner product on M by

(1.6) $$[u, v] = (Au, v).$$

Then the operator $\mathcal{A} = BA$ is symmetric and positive definite with respect to $[\cdot, \cdot]$. That is

$$[\mathcal{A}u, v] = (ABAu, v) = (u, ABAv) = [u, \mathcal{A}v]$$

and

$$[\mathcal{A}v, v] = (ABAv, v) = (BAv, Av) > 0, \text{ if } v \neq 0.$$

We now examine the convergence properties of (1.3) for general SPD operators B. Let $\{\psi_i\}$ and $\{\mu_i\}$ denote the eigenvectors and eigenvalues of $\mathcal{A}$, with $\mu_1 \leq \mu_2 \leq \cdots \leq \mu_N$. That is, $\mathcal{A}\psi_i = \mu_i \psi_i$, and we may assume that

$$[\psi_i, \psi_j] = \delta_{ij}.$$

3

Then, since

$$e^n = (I - BA)^n e^0 = (I - \mathcal{A})^n e^0$$

and

$$e^0 = \sum_{i=1}^{N} [e^0, \psi_i]\psi_i,$$

we see that

$$e^n = \sum_{i=1}^{N} [e^0, \psi_i](1 - \mu_i)^n \psi_i.$$

Hence $e^n \to 0$ for all e^0 if and only if

$$|1 - \mu_i| < 1$$

or

$$0 < \mu_i < 2.$$

Clearly if $\mu_N < 2$, then $-1 < 1 - \mu_N \le 1 - \mu_1 < 1$. Hence

$$\rho \equiv \max(|1 - \mu_1|, |1 - \mu_N|) < 1$$

and

$$[e^n, e^n]^{1/2} \le \rho^n [e^0, e^0]^{1/2}.$$

Again, if we find numbers c_0 and c_1 such that

(1.7) $$c_0[\varphi, \varphi] \le [\mathcal{A}\varphi, \varphi] \le c_1[\varphi, \varphi],$$

then

$$0 < c_0 \le \mu_1 \le \frac{[\mathcal{A}\varphi, \varphi]}{[\varphi, \varphi]} \le \mu_N \le c_1.$$

Let $\varphi = A^{-1}\psi$. Then (1.7) is the same as

(1.8) $$c_0(A^{-1}\psi, \psi) \le (B\psi, \psi) \le c_1(A^{-1}\psi, \psi).$$

Consequently, if c_1 and c_0 are close to 1, then ρ is small. Note if $B = A^{-1}$, then $c_0 = c_1 = 1$ and $x^1 = x^0 - A^{-1}(Ax^0 - f) = A^{-1}f = x$ and the process converges in one step.

4

Eventually we want to construct operators B so that

1) $1 \le K(BA)$ is "small" (near 1).

2) B is easy to apply (its action is inexpensive to compute).

Ideally we would like to have the action of B cost about the same as the action of A, (not A^{-1}). Such an operator B is called a "good preconditioner for A". Any SPD operator B may be considered to be a preconditioner.

General linear iterative processes

A general consistent linear iterative process for the solution of $Ax = f$ may be defined as follows. For f and u^0 given,

$$(1.9) \qquad\qquad u^{n+1} = \mathcal{I}(u^n, f)$$

with the following properties:

a) (linearity) $\mathcal{I}(u, f) + \alpha \mathcal{I}(v, g) = \mathcal{I}(u + \alpha v, f + \alpha g)$

and

b) (consistency) for any ϕ, $\phi = \mathcal{I}(\phi, A\phi)$.

Now it follows immediately that $\phi \to \mathcal{I}(\phi, 0)$ and $f \to \mathcal{I}(0, f)$ are linear transformations. Set $Bf = \mathcal{I}(0, f)$. By consistency and linearity

$$(1.10) \qquad\qquad \phi = \mathcal{I}(0, A\phi) + \mathcal{I}(\phi, 0).$$

Hence (1.9) may be written

$$u^{n+1} = \mathcal{I}(0, f) + \mathcal{I}(u^n, 0) = u^n + \mathcal{I}(0, f - Au^n).$$

Hence

$$u^{n+1} = u^n - B(Au^n - f).$$

Thus we see that (1.3) is, in fact, the general form of a consistent linear iterative process. If B is symmetric, then the process is said to be symmetric.

Now convergence of the process means that the sequence defined by

$$w^{n+1} = \mathcal{I}(w^n, 0)$$

tends to zero for any choice of w^0. This means that the spectral radius of the linear transformation defined by $\phi \rightarrow \mathcal{I}(\phi, 0)$ is less than one. Now (1.10) may be written as

$$\mathcal{I}(\phi, 0) = \phi - \mathcal{I}(0, A\phi) = (I - BA)\phi.$$

Hence the condition for convergence is that $\sigma(I - BA)$ lies in the open unit disk in the complex plane.

The following observation is important. The action of B on a given f is the same as the result of one step in (1.9) with $u^0 = 0$. Thus we see that one step of a linear iterative process with $u^0 = 0$ gives rise to a preconditioner if the process is symmetric and convergent; i.e. with $Bf = \mathcal{I}(0, f)$, the spectral radius of $I - BA$ is equal to a nonnegative number $\delta < 1$. Thus we have

$$(1 - \delta)(A^{-1}u, u) \le (Bu, u) \le (1 + \delta)(A^{-1}u, u).$$

Conversely, if an SPD operator is given and $c_1 < 2$ in (1.8) then (1.3) defines a consistent, convergent linear iterative process.

Now to analyze such a process we may always study the operator $I - BA$ associated with (1.3) or (1.9). Note that it is symmetric with respect to $(A\cdot, \cdot) = [\cdot, \cdot]$ and therefore if the process is convergent, $\sigma(I - BA) \subset (-1, 1)$. Hence there exist numbers a_0 and a_1 in $[0, 1)$ such that

$$-a_0(Au, u) \le (A(I - BA)u, u) \le a_1(Au, u)$$

or

$$(1 - a_1)(Au, u) \le (ABAu, u) \le (1 + a_0)(Au, u).$$

This is the same as

$$(1 - a_1)(A^{-1}u, u) \le (Bu, u) \le (1 + a_0)(A^{-1}u, u),$$

which is equivalent to

$$(1 - a_1)(B^{-1}u, u) \le (Au, u) \le (1 + a_0)(B^{-1}u, u)$$

6

(see Lemma 3.2). Note that

$$0 < 1 - a_1 \text{ and } 1 + a_0 < 2.$$

Now assume that we know constants c_0 and c_1 which satisfy

$$c_0(B^{-1}u, u) \le (Au, u) \le c_1(B^{-1}u, u)$$

with $0 < c_0 \le 1, 1 \le c_1 < 2$. Set $a_1 = 1 - c_0$, $a_0 = c_1 - 1$. Then $0 \le a_1, a_0 < 1$,

$$-a_0(Au, u) \le (A(I - BA)u, u) \le a_1(Au, u)$$

and

$$|||I - BA||| = \rho \le \max(a_0, a_1) < 1,$$

where $||| \cdot |||^2 = (A \cdot, \cdot)$ and $|||I - BA|||$ is the corresponding operator norm. Consequently, estimates for c_0 and c_1 yield convergence estimates for the linear iteration. If, in fact, $a_0 = a_1 = a$ and $K = \frac{1+a}{1-a}$, then $\rho \le a = \frac{K-1}{K+1} < 1$.

We finish the first section by briefly discussing the important conjugate gradient algorithm. The conjugate gradient (CG) algorithm is a non–linear iteration which may be described as follows. The nth CG iterate is determined by means of a certain projection onto the Krylov subspace V_n of Appendix A. This is made precise in Appendix A. As will be seen, if no round–off errors are introduced into the computation, then CG obtains the solution in at most N steps. This property is not the really important one. What is of great importance, however, is the behavior of the error with respect to n, the number of iterations. We state here this estimate with starting guess zero for the sake of discussion. The estimate of the error is

$$(A(x - y_n), (x - y_n)) \le 2\rho^{2n}(Ax, x)$$

where

$$\rho = \frac{K(A)^{1/2} - 1}{K(A)^{1/2} + 1} = \frac{K(A) - 1}{(K(A)^{1/2} + 1)^2} \le \frac{K(A) - 1}{K(A) + 1}.$$

7

We see from this that in the case that N is very large, the error may be acceptably small for values of n much smaller than N. Hence, even though, in principle, CG is a direct method, its power rests in considering it as an iterative method.

Of course, the above observations are important in practice only if the computation of the nth iterate is reasonable. As is discussed in detail in Appendix A, the nth iterate in CG is given by means of a two term recurrence relation and thus the computational properties are very favorable.

Finally, we mention here that effective use can be made of preconditioning in the CG algorithm. Details are in Appendix A.

Bibliographical Notes

The linear iterative method (1.3) is a special case of the classical Picard method of successive approximations which converges to a unique fixed point when the mapping is a contraction. For more discussion of iterative methods cf. [94] and [154]. The presentation here is meant to emphasize the importance of preconditioning.

2. Model problems and Sobolev spaces

Consider, for $\Omega \subset R^2$,

$$(2.1) \qquad -\Delta u = f, \text{ in } \Omega, \ \Delta = \frac{\partial^2}{\partial x^2} + \frac{\partial^2}{\partial y^2}$$

$$u = 0 \text{ on } \partial\Omega.$$

This is the well–known Dirichlet Problem. Set

$$(v, w) = \int_\Omega vw \, dx \, dy$$

and

$$D(v, w) = \int_\Omega \left(\frac{\partial v}{\partial x} \frac{\partial w}{\partial x} + \frac{\partial v}{\partial y} \frac{\partial w}{\partial y} \right) dx \, dy.$$

Denote by $C_0^\infty(\Omega)$ the space of infinitely differentiable functions with compact support in Ω. For $\varphi \in C_0^\infty$ we have (formally)

$$(f, \varphi) = (-\Delta u, \varphi) = D(u, \varphi).$$

Let $H^1(\Omega)$ be the usual Sobolev space with norm $\|v\|_1^2 = D(v, v) + (v, v)$. Let $H_0^1(\Omega)$ be the closure of $C_0^\infty(\Omega)$ with respect to $\|\cdot\|_1^2$. The Poincaré inequality says that there is a constant $C > 0$ such that

$$(2.2) \qquad (v, v) \equiv \|v\|_0^2 \leq CD(v, v), \text{ for all } v \in H_0^1(\Omega).$$

Hence we may take $D(\cdot, \cdot)^{1/2}$ to be the norm on $H_0^1(\Omega)$. This changes the Hilbert space structure.

<u>Weak solution of (2.1)</u>. A natural description of solutions of (2.1) is the following variational or weak formulation of (2.1): Find $u \in H_0^1(\Omega)$ such that

$$(2.3) \qquad D(u, \varphi) = (f, \varphi), \text{ for all } \varphi \in H_0^1(\Omega).$$

9

Existence and uniqueness will follow from the Poincaré inequality, (2.2), and the Riesz representation theorem which we now state .

The Riesz representation theorem: Let H be a Hilbert space and $F(\cdot)$ be a bounded linear functional on H; i.e., there exists a constant $C(F)$ such that

$$|F(\varphi)| \leq C(F)\|\varphi\|, \text{ for all } \varphi \in H.$$

Then there exists a unique $u_F \in H$ such that

$$((u_F, \varphi)) = F(\varphi), \text{ for all } \varphi \in H,$$

where $((\cdot, \cdot))$ is the inner product in H.

Using this theorem, we obtain existence and uniqueness of the solution of (2.3) as follows. Set

$$F(\varphi) = (f, \varphi) = \int_\Omega f\varphi dxdy, \text{ for } \varphi \in H_0^1(\Omega).$$

Then

$$|F(\varphi)| \leq \|f\|_0 \|\varphi\|_0 \leq C\|f\|_0 D^{1/2}(\varphi, \varphi)$$

by (2.2).

Now $D(\varphi, \psi) = \int_\Omega \nabla\varphi \cdot \nabla\psi dxdy$ is the inner product on $H_0^1(\Omega)$. Hence F is a bounded linear functional on $H_0^1(\Omega)$ so that there is a unique $u_F \in H_0^1(\Omega)$ such that

$$D(u_F, \varphi) = (f, \varphi), \text{ for all } \varphi \in H_0^1(\Omega).$$

We will also want to refer to one of the simplest regularity results for the solution u of (2.3). If Ω has a smooth boundary (cf. [110]) or is a convex polygon (cf. [96]), then there is a constant C such that

(2.4)
$$\|u\|_2 \leq C\|f\|_0.$$

Here

$$\|u\|_2^2 = \sum_{|\alpha| \leq 2} \int_\Omega |D^\alpha u|^2 dxdy,$$

where

$$D^\alpha u = \frac{\partial^{|\alpha|} u}{\partial^{\alpha_1} x \, \partial^{\alpha_2} y},$$

$\alpha = (\alpha_1, \alpha_2)$ is a multi-index with $|\alpha| = \alpha_1 + \alpha_2$; e.g., $D^{(0,0)} = u$, $D^{(1,0)} u = \frac{\partial u}{\partial x}$, etc.

We now can introduce the Galerkin approximation in a finite dimensional subspace M of $H_0^1(\Omega)$. It is defined to be the solution of the following problem: Find $U \in M$ such that

(2.5) $$D(U, \chi) = (f, \chi), \text{ for all } \chi \in M.$$

We note that (2.5) always has a unique solution. In fact, if $f = 0$ then $D(U, U) = 0$ which implies that $U = 0$ which, in turn, implies existence and uniqueness in the finite dimensional case.

Now the simplest finite element method is obtained by making a special choice of M. For the purpose of our discussion of the multigrid method let Ω be a convex polygon and let τ_1 be a (coarse) triangulation of Ω. We define M_1 to be the space of continuous piecewise linear functions on τ_1 which vanish on $\partial\Omega$. Let h_1 be the length of the side of maximum length in τ_1.

If the triangulation τ_k and the space M_k have been defined, define τ_{k+1} by subdividing each triangle of τ_k into four triangles by joining the midpoints of the sides. Define M_{k+1} analogously relative to τ_{k+1}. Thus $M_1 \subset M_2 \subset \cdots \subset M_J = M \subset H_0^1(\Omega)$. We define $h_{k+1} = \frac{1}{2} h_k$ and $h = h_J$. Then the finite element approximation is the following: Find $U_h \in M$ such that

(2.6) $$D(U_h, \chi) = (f, \chi), \text{ for all } \chi \in M.$$

In order to estimate the error $U_h - u$ we note the following well-known approximation estimate, cf. [65] . For $v \in H^r(\Omega)$, $r = 1$ and 2, there exists $\chi \in M$ such that

(2.7) $$\|v - \chi\|_0^2 + h^2 D(v - \chi, v - \chi) \leq C h^{2r} \|v\|_r^2$$

11

with C independent of v and h. From this we easily derive an error estimate for (2.6). From (2.3) and (2.6)

$$D(U_h, \chi) = (f, \chi) = D(u, \chi), \text{ for all } \chi \in M.$$

Hence

$$D(U_h - u, U_h - u) = D(U_h - u, \chi - u), \text{ for all } \chi \in M.$$

This implies that, for $r = 1$ or 2 and $u \in H^r(\Omega)$,

$$\|U_h - u\|_1^2 \leq cD(U_h - u, U_h - u)$$
$$\leq cD(\chi - u, \chi - u)$$
$$\leq ch^{2r-2}\|u\|_r^2.$$

Take $r = 2$. Then

$$\|U_h - u\|_1^2 \leq ch^2\|u\|_2^2$$

or

$$\|U_h - u\|_1 \leq ch\|u\|_2.$$

We next want to show that the convergence is better with respect to the weaker norm $\|U_h - u\|_0$. We show this by a standard duality argument often called the "Aubin–Nitsche trick". This is done as follows. For Ω, a convex polygonal domain, we use (2.4). That is, there is a constant $c > 0$ such that for $v \in H_0^1(\Omega) \cap H^2(\Omega)$

(2.8) $$\|v\|_2 \leq c\|\Delta v\|_0.$$

Write, using (2.3),

$$\|U_h - u\|_0^2 = D(U_h - u, v)$$

where $-\Delta v = U_h - u$ in Ω and $v = 0$ on $\partial\Omega$. Then, for any $\chi \in M$,

$$\|U_h - u\|_0^2 = D(U_h - u, v)$$
$$= D(U_h - u, v - \chi)$$
$$\leq D(U_h - u, U_h - u)^{1/2}D(v - \chi, v - \chi)^{1/2}$$
$$\leq Ch\|v\|_2 D(U_h - u, U_h - u)^{1/2}$$
$$\leq Ch\|U_h - u\|_0 D(U_h - u, U_h - u)^{1/2},$$

12

where we have used the Cauchy–Schwarz inequality, (2.7) and (2.8). Hence

$$\|U_h - u\|_0 \leq ChD(U_h - u, U_h - u)^{1/2}$$

$$\leq Ch^2 \|u\|_2.$$

Thus we get a faster rate of convergence in $L^2(\Omega)$ than in $H_0^1(\Omega)$.

We want to discuss now the computation of U_h and in this connection some properties of the linear system determining U_h. Order the interior vertices of τ_J in some way from 1 to N_J, where N_J is the dimension of M_J. Let $\varphi_i \in M$ be such that

$$\varphi_i = \begin{cases} 1 & \text{at } x_i \\ 0 & \text{at } x_j \neq x_i. \end{cases}$$

Then clearly, since any continuous, piecewise linear function is determined by the values at the vertices,

$$U_h = \sum_{i=1}^{N_J} \alpha_i \varphi_i.$$

Hence we get, for $j = 1, 2, \ldots, N_J$

$$\sum_{i=1}^{N_J} \alpha_i D(\varphi_i, \varphi_j) = (f, \varphi_j).$$

It is easy to show that if

$$\chi = \sum_{i=1}^{N_J} c_i \varphi_i$$

then

$$\|\chi\|_0^2 \approx \|\chi\|_{0,h}^2 \equiv h^2 \sum_{i=1}^{N_J} c_i^2 = h^2 \sum_{i=1}^{N_J} \chi^2(x_i).$$

The notation $\approx$ means equivalence of norms with constants of equivalence independent of h. Hence, by the Poincaré inequality, (2.2),

$$Ch^2 \sum_{i=1}^{N_J} c_i^2 \leq \sum_{i,j=1}^{N_J} c_i c_j D(\varphi_i, \varphi_j),$$

with C a constant which is independent of h and χ. This means that the smallest eigenvalue λ_{min} of the matrix $D(\varphi_i, \varphi_j)$ satisfies

(2.9) $$\lambda_{min} \geq Ch^2.$$

Here and in the following, the notation C and c will be used in a standard way to denote generic constants independent of h and any particular functions involved. Now it is an exercise to check that

$$D(\chi, \chi) = \sum_{i,j=1}^{N_J} c_i c_j D(\varphi_i, \varphi_j)$$

$$\leq C \sum_{i=1}^{N_J} c_i^2 D(\varphi_i, \varphi_i)$$

$$\leq C \sum_{i=1}^{N_J} c_i^2.$$

Hence $\lambda_{max} \leq C$ and, using (2.9), this means that the condition number $K(A) = \lambda_{max}/\lambda_{min}$ satisfies

$$K(A) \leq Ch^{-2},$$

where A is the matrix with entries $D(\varphi_i, \varphi_j)$.

We shall show that this estimate is sharp in the sense that $h^2 K(A)$ is bounded below also by a constant independent of h for small h. Now

$$\lambda_{min} h^{-2} \leq \frac{D(\chi, \chi)}{\|\chi\|_{0,h}^2} \leq \lambda_{max} h^{-2}.$$

Take $\chi = \varphi_i$ for some i. Then

$$\frac{D(\varphi_i, \varphi_i)}{h^2} = \frac{c}{h^2} \leq \lambda_{max} h^{-2}$$

and therefore

$$\lambda_{max} \geq c > 0.$$

To obtain an upper bound for λ_{min}, let u_1 be an eigenvector corresponding to the lowest eigenvalue μ_1 in the fixed membrane problem

$$-\Delta u_1 = \mu_1 u_1 \quad \text{in } \Omega$$

$$u_1 = 0 \quad \text{on } \partial\Omega.$$

Then choose χ_h such that

$$\|u_1 - \chi_h\|_0^2 + h^2 D(u_1 - \chi_h, u_1 - \chi_h) \le \hat{c}^2 h^2 D(u_1, u_1).$$

Hence

$$D^{1/2}(\chi_h, \chi_h) \le D^{1/2}(u_1, u_1) + D^{1/2}(u_1 - \chi_h, u_1 - \chi_h)$$

$$\le (1 + \hat{c}) D^{1/2}(u_1, u_1)$$

and

$$\|\chi_h\|_{0,h} \ge c\|\chi_h\|_0$$

$$\ge c(\|u_1\|_0 - \|u_1 - \chi_h\|_0)$$

$$\ge c(\frac{1}{\mu_1^{1/2}} - \hat{c}h) D^{1/2}(u_1, u_1).$$

Consequently, for h such that $\mu_1^{1/2} \hat{c} h < 1/2$ we have

$$\|\chi_h\|_{0,h} \ge \tilde{c} D^{1/2}(u_1, u_1).$$

Thus

$$h^{-2} \lambda_{min} \le \frac{(1 + \hat{c})^2}{\tilde{c}^2}.$$

Hence $\lambda_{min}^{-1} \ge \frac{\tilde{c}^2}{(1+\hat{c})^2} h^{-2}$ and it follows that $\frac{c\tilde{c}^2}{(1+\hat{c})^2} \le h^2 K(A)$ and therefore the bound is sharp. This means that the problem is rather ill conditioned for h small. The simple linear iterative method

$$x^{n+1} = x^n - \lambda_{max}^{-1}(Ax^n - f)$$

is such that the ith frequency of the error is reduced by

$$\rho_i = (1 - \lambda_{max}^{-1} \nu_i),$$

with ν_i an eigenvalue of A. For ν_i an eigenvalue close to λ_{max}, the reduction is $1 - \nu_i/\lambda_{max} < 1$ which is quite good. But for ν_i near $\lambda_{min} = ch^2$, $\rho_i = 1 - ch^2$ which is near 1 and hence very bad for h very small.

Let us turn our attention again to (2.6). Anticipating a more general situation which we will be considering later, set $A(u,v) \equiv D(u,v)$, $u,v \in M = M_J$. Now recall that $M_{J-1} \subset M_J$, and in general $M_1 \subset M_2 \subset \ldots \subset M_J = M$. Define the linear operator $A_k : M_k \to M_k$ by

(2.10)
$$(A_k v, \varphi) = A(v, \varphi), \text{ for all } \varphi \in M_k.$$

Clearly, given v, A_k makes sense. Now A_k is symmetric and positive definite. Denote by λ_k the largest eigenvalue of A_k.

A very simple iteration can be written as

(2.11)
$$x^{n+1} = x^n - \overline{\lambda}_J^{-1}(A_J x^n - f_J),$$

where $f_J \in M$ is defined by

$$(f_J, \varphi) = (f, \varphi), \text{ for all } \quad \varphi \in M$$

and

$$\lambda_J \leq \overline{\lambda}_J \leq \bar{c}\lambda_J.$$

Note that

$$c_0 \leq \frac{(A_J\varphi, \varphi)}{\|\varphi\|_0^2} \leq c_1 h^{-2}.$$

Thus

$$K(A_J) \approx ch^{-2}.$$

The iteration (2.11), in general, has very poor convergence properties. However, it does have the property that the high frequecy components of the error are damped as discussed in Section 1. We shall consider alternatives to this simple iteration in which we take advantage of this "smoothing" property.

We want to modify the method defined by (2.11). To this end define the projectors

$$P_k : M_J \to M_k \text{ and } Q_k : M_J \to M_k$$

16

by

$$A(P_k v, \varphi) = A(v, \varphi), \text{ for all } \varphi \in M_k,$$

and

$$(Q_k v, \varphi) = (v, \varphi), \text{ for all } \varphi \in M_k.$$

Clearly P_k and Q_k are well defined.

We now define a new iteration as follows.

Two level algorithm:

0) Let u_0 be given

For u_i "approximating" u, the solution of $A_J u = f_J$, define u_{i+1} as follows:

1) Set $x_1 = u_i - \overline{\lambda}_J^{-1}(A_J u_i - f_J)$ (smooth)

2) $x_2 = x_1 - q$, where $A_{J-1} q = Q_{J-1}(A_J x_1 - f_J)$ (correct)

3) $u_{i+1} = x_2 - \overline{\lambda}_J^{-1}(A_J x_2 - f_J)$ (smooth).

Clearly this is a consistent linear iterative process and the operator $\overline{\lambda}_J^{-1} I$ is our first example of a smoothing operator.

We end this section by analyzing the convergence of the two level algorithm. For this purpose let $|||V||| = A(V, V)^{1/2}$. We will prove the following.

THEOREM 2.1. *Set* $e_i = u - u_i$. *Then*

$$|||e_{i+1}||| \le \delta |||e_i|||$$

where $\delta < 1$ *and independent of* u_i, f_J *and* h.

Proof: We first see that $Q_{k-1} A_k = A_{k-1} P_{k-1}$. In fact, for $v \in M_k$, $\chi \in M_{k-1}$,

$$(Q_{k-1} A_k v, \chi) = (A_k v, \chi) = A(v, \chi) = A(P_{k-1} v, \chi) = (A_{k-1} P_{k-1} v, \chi).$$

Let $E_j = u - x_j$. By consistency

$$E_1 = (I - \overline{\lambda}_J^{-1} A_J) e_i$$

and

$$q = -A_{J-1}^{-1} Q_{J-1} A_J E_1 = -P_{J-1} E_1$$

so that $E_2 = u - x_2 = u - x_1 + q = (I - P_{J-1})E_1$, i.e.,

$$E_2 = (I - P_{J-1})E_1.$$

Again, by consistency,

$$\begin{aligned}
e_{i+1} = E_3 &= (I - \overline{\lambda}_J^{-1} A_J)E_2 \\
&= (I - \overline{\lambda}_J^{-1} A_J)(I - P_{J-1})(I - \overline{\lambda}_J^{-1} A_J)e_i \\
&\equiv K_{mg}e_i.
\end{aligned}$$

Thus

$$\||e_{i+1}\|| \leq \||K_{mg}\|| \, \||e_i\||,$$

where $\||K_{mg}\||$ is the operator norm of K_{mg}. Note that K_{mg} is symmetric in $A(\cdot,\cdot)$.

Hence

$$\begin{aligned}
\||K_{mg}\|| &= \sup_{v \in M_J} \frac{|A(K_{mg}v,v)|}{\||v\||^2} \\
&= \sup_{v \in M_J} \frac{\||(I - P_{J-1})(I - \overline{\lambda}_J^{-1} A_J)v\||^2}{\||v\||^2} \\
&= \||(I - P_{J-1})(I - \overline{\lambda}_J^{-1} A_J)\||^2 \\
&= \||(I - \overline{\lambda}_J^{-1} A_J)(I - P_{J-1})\||^2.
\end{aligned}$$

Set $\tilde{\chi} = (I - P_{J-1})\chi$, for $\chi \in M_J$. Then $A(\tilde{\chi}, \theta) = 0$, for all $\theta \in M_{J-1}$. Now

$$A(\tilde{\chi}, \tilde{\chi}) = A(\tilde{\chi}, \tilde{\chi} - \theta) = (A_J\tilde{\chi}, \tilde{\chi} - \theta) \leq \|A_J\tilde{\chi}\|_0 \|\tilde{\chi} - \theta\|_0.$$

Using the approximation property (2.7),

$$A(\tilde{\chi}, \tilde{\chi}) \leq Ch_J \|A_J\tilde{\chi}\|_0 A(\tilde{\chi}, \tilde{\chi})^{1/2}.$$

Thus

$$\||\tilde{\chi}\|| \leq Ch_J \|A_J\tilde{\chi}\|_0,$$

and, since $h_J^2 \leq c\lambda_J^{-1}$, it follows that

$$\||\tilde{\chi}\||^2 \leq \tilde{C}\lambda_J^{-1} \|A_J\tilde{\chi}\|_0^2.$$

18

Now $I - \overline{\lambda}_J^{-1} A_J$ is symmetric in $A(\cdot, \cdot)$ and $\sigma(I - \overline{\lambda}_J^{-1} A_J) \subseteq [0, 1)$. Hence

$$(2.12) \qquad |||(I - \overline{\lambda}_J^{-1} A_J)\tilde{\chi}|||^2 \leq A((I - \overline{\lambda}_J^{-1} A_J)\tilde{\chi}, \tilde{\chi})$$
$$= |||\tilde{\chi}|||^2 - \overline{\lambda}_J^{-1} \|A_J \tilde{\chi}\|_0^2$$
$$\leq (1 - 1/\tilde{C})|||\tilde{\chi}|||^2$$
$$\leq (1 - 1/\tilde{C})|||\chi|||^2.$$

Therefore

$$|||e_{i+1}||| \leq (1 - 1/\tilde{C})|||e_i||| \equiv \delta|||e_i|||.$$

If we omit Step 1 then

$$e_{i+1} = (I - \overline{\lambda}_J^{-1} A_J)(I - P_{J-1})e_i$$

and hence

$$|||e_{i+1}||| = |||(I - \overline{\lambda}_J^{-1} A_J)(I - P_{J-1})e_i|||$$
$$\leq |||(I - \overline{\lambda}_J^{-1} A_J)(I - P_{J-1})||| \; |||e_i|||$$
$$\leq \delta^{1/2}|||e_i|||.$$

The same result is true if we omit instead Step 3. One obvious advantage of the symmetric algorithm is that it can be used to define a preconditioner.

Why have we gained by the two level scheme? In this example

$$\dim M_{J-1} \approx \frac{1}{4} \dim M_J.$$

Hence A_{J-1} is much cheaper to invert than is A_J. Since A_{J-1} must be "inverted" at each step of the iteration, this scheme may not be a good one to use in practice. Its introduction is meant to be a point of departure for the development of more general multilevel algorithms. As will be seen in the subsequent sections, the generalizations of such a scheme to an arbitrary number of levels give rise to powerful tools for the efficient solution of (1.1).

Bibliographical Notes

Estimates of the form of (2.7) are well known (cf. [65]).

The duality argument, sometimes called the "Aubin–Nitsche trick", was given independently by Aubin [4] and Nitsche [134]. It has become a standard technique in the error analysis of the finite element method (cf. [65]). Regularity estimates such as (2.8) for convex polygonal domains in the plane may be found in [96]. Other estimates of this type for general plane domains with polygonal boundaries may be found in [69].

The argument for the two level algorithm was first given by Mandel [114] . See also the thesis of Xu [164]. This argument does not require regularity of the continuous problem.

3. Abstract Multilevel Algorithms

In the previous section we introduced a two level algorithm for a model problem and proved, under certain natural conditions, its uniform convergence. Keeping in mind the example of Section 2, we shall introduce in this section an abstract setting in which the levels are defined by a nested set of subspaces of the space M where the solution of (1.1) lies. First, we shall present one of the simplest multilevel algorithms, generalizing the two level algorithm of Section 2. Later in the section more general algorithms are introduced within the present framework. They will allow for more smoothing or correction steps. Various abstract conditions will be introduced along the way and a number of theorems concerning the convergence properties of the algorithms will be proved. The meanings of the conditions will be illustrated by specific examples in Sections 6, 7 and 8.

We start with a finite dimensional space M equipped with two inner products $(\cdot, \cdot)$ and $A(\cdot, \cdot)$ with corresponding norms $\| \cdot \|$ and $\| \| \cdot \| \|$. Now set $M = M_J$ and suppose that we have subspaces M_k with

$$M_1 \subset M_2 \subset \cdots \subset M_J \equiv M.$$

Recall that we defined in (2.10) the linear operator $A_k : M_k \to M_k$ by

$$(A_k \psi, \theta) = A(\psi, \theta), \text{ for all } \psi, \theta \in M_k.$$

Furthermore, the projectors $P_k : M_J \to M_k$ and $Q_k : M_J \to M_k$ were defined in Section 2 by

$$A(P_k u, v) = A(u, v)$$

and

$$(Q_k u, v) = (u, v),$$

for all $u \in M_J$ and all $v \in M_k$. In the model problem of Section 2, Q_k is the L_2–projection and P_k is the so–called elliptic projection.

We also need to introduce now a generic linear smoothing operator $R_k : M_k \to M_k$, but we will always take $R_1 = A_1^{-1}$. We do not assume, in general, that the operator R_k is symmetric. Let R_k^t denote the adjoint of R_k with respect to $(\cdot, \cdot)$; i.e.,

$$(R_k^t \psi, \theta) = (\psi, R_k \theta), \text{ for all } \psi, \theta \in M_k.$$

Examples of useful smoothing operators are given in Section 5.

V–cycle algorithm: We want to define the operator $B = B_J$ which can be used in a linear iterative process of the type (1.3). For this purpose define an operator $B_k : M_k \to M_k$ inductively as follows.

Algorithm I:

0) $B_1 = A_1^{-1}$ (solve)

For $k > 1$, B_k is defined in terms of B_{k-1} as follows. Let $g \in M_k$

1) $x_1 = R_k^t g$ (smooth)

2) $x_2 = x_1 - q$, where $q = B_{k-1} Q_{k-1}(A_k x_1 - g)$ (correct)

3) $B_k g = x_2 - R_k(A_k x_2 - g)$ (smooth).

Notice that for $k = 2$, $u_0 = 0$ and $R_2 = \overline{\lambda_2}^{-1} I$ this is the 2–level algorithm of Section 2.

We will see that the operators B_k are symmetric with respect to $(\cdot, \cdot)$ and later we will see that, under reasonable conditions, is also positive definite. Thus, as was discussed in Section 1, $B = B_J$ can be used to define a linear iterative process or a preconditioner. In any case the relevant properties may be deduced from the study of the related operators $I - B_k A_k$.

To this end we first derive a recurrence relation for the error operator $I - B_k A_k$. Set $g = A_k x$ in M_k, $K_k = I - R_k A_k$ on M_k and

(3.1) $$x - x_1 = K_k^* x,$$

22

where K_k^* is the $A(\cdot, \cdot)$-adjoint of K_k and is given by

$$K_k^* = I - R_k^t A_k.$$

In fact, for ψ and θ in M_k,

$$
\begin{aligned}
A(K_k\psi, \theta) &= A((I - R_k A_k)\psi, \theta) \\
&= ((I - R_k A_k)\psi, A_k\theta) \\
&= (A_k\psi, (I - R_k^t A_k)\theta) \\
&= A(\psi, K_k^*\theta).
\end{aligned}
$$

Now

$$q = -B_{k-1}Q_{k-1}A_k(x - x_1)$$

so that

$$x - x_2 = x - x_1 - B_{k-1}Q_{k-1}A_k(x - x_1).$$

Recall the relationship essentially proved in the proof of Theorem 2.1, namely that

$$Q_{k-1}A_\ell = A_{k-1}P_{k-1} \text{ on } M_\ell,$$

for $\ell \geq k$. Hence

(3.2) $$x - x_2 = (I - B_{k-1}A_{k-1}P_{k-1})(x - x_1) = (I - B_{k-1}A_{k-1}P_{k-1})K_k^* x$$

and

$$
\begin{aligned}
(I - B_k A_k)x &= x - x_2 - R_k A_k(x - x_2) \\
&= K_k(x - x_2) \\
&= K_k[I - B_{k-1}A_{k-1}P_{k-1}]K_k^* x \\
&= K_k[(I - P_{k-1}) + (I - B_{k-1}A_{k-1})P_{k-1}]K_k^* x.
\end{aligned}
$$

Thus, since $x \in M_k$ is arbitrary

(3.3) $$I - B_k A_k = K_k[(I - P_{k-1}) + (I - B_{k-1}A_{k-1})P_{k-1}]K_k^*.$$

23

It follows by induction that $B_k A_k$ is symmetric with respect to $A(\cdot, \cdot)$ and hence B_k is symmetric with respect to $(\cdot, \cdot)$. This is an important two level recurrence relation.

We next want to show, using (3.3), that the operator $I - B_J A_J$ may be expressed as a product of operators, each of which is defined on all of M_J. To see this we extend K_k to all of M_J by

$$K_k = I - R_k A_k P_k \equiv I - T_k.$$

Thus

$$K_k^* = I - R_k^t A_k P_k = I - T_k^*.$$

We call the extension again K_k since there should be no confusion. From (3.3) we have

$$I - B_k A_k P_k = (I - P_k) + (I - B_k A_k) P_k = I - P_k + K_k [I - B_{k-1} A_{k-1} P_{k-1}] K_k^* P_k.$$

Now $R_k^t : M_k \to M_k$ so that

$$(I - P_k)(I - T_k^*) = I - P_k.$$

Also $(I - T_k)(I - P_k) = I - P_k$ and $(I - P_k)^2 = I - P_k$. Hence

$$I - P_k = (I - T_k)(I - P_k)(I - T_k^*) = K_k(I - P_k) K_k^*.$$

Therefore

$$I - B_k A_k P_k = (I - T_k)[I - P_k + P_k - B_{k-1} A_{k-1} P_{k-1}](I - T_k^*)$$

or

$$I - B_k A_k P_k = (I - T_k)(I - B_{k-1} A_{k-1} P_{k-1})(I - T_k^*).$$

Hence

$$(3.4) \quad I - B_J A_J = (I - T_J)(I - T_{J-1}) \ldots (I - T_2)(I - P_1)(I - T_2^*) \ldots (I - T_J^*).$$

24

Thus the error operator $I - B_J A_J$ in the multigrid V–cycle is a product of operators each of which is defined on all of M_J.

Remark: Assumptions concerning the operator R_k will be made later. Section 5 is devoted to the construction of some useful classes of smoothers satisfying these assumptions.

In order to introduce some conditions needed to study the multigrid methods we shall define and briefly study a special "additive" preconditioner. This operator will be defined by making a special choice of T_k, or what is the same, R_k. This operator is introduced here purely for theoretical purposes. Some alternatives to this operator, which are useful in some examples, may be found in [45].

Set $\tilde{T}_1 = P_1$ and $\tilde{T}_k = \frac{1}{\lambda_k} A_k P_k = \frac{1}{\lambda_k} Q_k A_J$, $k > 1$, where we recall that λ_k is the largest eigenvalue of A_k. This corresponds to the choice

$$\tilde{R}_k = \frac{1}{\lambda_k} I, \ k > 1, \ \tilde{R}_1 = A_1^{-1}.$$

Set

$$\tilde{B}^a = \sum_{k=1}^{J} \tilde{R}_k Q_k$$

so that, setting $A \equiv A_J$,

$$\tilde{B}^a A \equiv \tilde{B}^a A_J = \sum_{k=1}^{J} \tilde{T}_k.$$

Now $\tilde{B}^a A$ is symmetric in $A(\cdot, \cdot)$ so its lowest eigenvalue can be estimated by finding a constant C_a such that

(3.5) $\qquad A(u,u) \le C_a A(\tilde{B}^a A u, u) = C_a \sum_{k=1}^{J} A(\tilde{T}_k u, u), \ \text{for all } u \in M.$

We shall subsequently take this as Condition A.1.

Let us digress a moment and show that in the case of our example in Section 2, i.e. the Dirichlet problem on a convex polygonal domain, (3.5) is easily proved.

Let $v \in M_k$, for $k \geq 2$. Then

$$A((I - P_{k-1})v, v) = ((I - P_{k-1})v, A_k v) \leq \|(I - P_{k-1})v\| \, \|A_k v\|.$$

We saw from the "Aubin–Nitsche trick" that, since Ω is convex,

$$\|(I - P_{k-1})v\| \leq C h_{k-1} A((I - P_{k-1})v, (I - P_{k-1})v)^{1/2}$$

$$\leq C \lambda_k^{-1/2} A((I - P_{k-1})v, v)^{1/2}.$$

Hence $A((I - P_{k-1})v, v) \leq C \lambda_k^{-1} \|A_k v\|^2$. Now for $u \in M_J$ set $v = P_k u$. Notice that

$$A(P_{k-1} P_k u, \varphi) = A(P_k u, P_{k-1}\varphi) = A(u, P_k P_{k-1}\varphi), \text{ for all } \varphi \in M.$$

Clearly $P_k P_{k-1} = P_{k-1}$ since $M_{k-1} \subset M_k$ and hence

$$A(P_{k-1} P_k u, \varphi) = A(u, P_{k-1}\varphi) = A(P_{k-1} u, \varphi), \text{ for all } \varphi \in M.$$

Thus $P_{k-1} P_k = P_{k-1} = P_k P_{k-1}$. Hence for $k \geq 2$,

$$(3.6) \qquad A((P_k - P_{k-1})u, u) = A((I - P_{k-1})P_k u, P_k u) \leq C \lambda_k^{-1} \|A_k P_k u\|^2.$$

Setting $P_0 = 0$ and summing, we get

$$A(u, u) = \sum_{k=1}^{J} A((P_k - P_{k-1})u, u)$$

$$\leq A(P_1 u, u) + C \sum_{k=2}^{J} \lambda_k^{-1} \|A_k P_k u\|^2.$$

Assume, without loss, that $C = C_a \geq 1$. Then we have shown that

$$(3.7) \qquad A(u, u) \leq C_a \left(A(P_1 u, u) + \sum_{k=2}^{J} \lambda_k^{-1} A(A_k P_k u, u) \right)$$

$$= C_a \sum_{k=1}^{J} A(\tilde{T}_k u, u).$$

26

Remark: The inequality (3.7) may be considered as a weak version of (3.6) in that (3.7) holds in many cases when (3.6) does not, for instance, when the domain is not convex so that the regularity property used above does not hold.

We thus formulate (3.5) as a condition.

A.1: There exists $C_a > 0$ such that

$$A(v,v) \leq C_a \sum_{k=1}^{J} A(\tilde{T}_k v, v), \text{ for all } v \in M.$$

In order to get a corresponding upper bound we shall assume another condition.

A.2: There exist $0 < \varepsilon < 1$ and $\tilde{C} > 0$ such that

$$A(\tilde{T}_k w, w) \leq (\tilde{C}\varepsilon^{k-\ell})^2 A(w,w),$$

for all $w \in M_\ell, \ell \leq k$.

This condition will be seen later to be closely related to a so-called inverse property of the subspaces. It is shown to hold in the case of the integral operator in Section 8, specifically (8.18). It may also be shown to be satisfied in the examples on second order elliptic problems. The proof in the second order elliptic case, though not contained in Section 6, may be found in [36].

The following lemma provides a tool useful in establishing A.1.

LEMMA 3.1. *Assume that we have two SPD operators $\mathcal{A}$ and A on M_J with corresponding forms $\mathcal{A}(\cdot,\cdot)$ and $A(\cdot,\cdot)$ on $M_J \times M_J$ and positive constants C_0 and C_1 such that*

$$(3.8) \qquad C_0 A(u,u) \leq \mathcal{A}(u,u) \leq C_1 A(u,u) \text{ for all } u \in M_J.$$

Then A.1 holds for A if and only if A.1 holds for $\mathcal{A}$.

We note first another simple but important lemma.

LEMMA 3.2. *Suppose $\mathcal{A}$ and A are two SPD operators. Then, for all $u \in M$,*

$$C_0(Au, u) \leq (\mathcal{A}u, u) \leq C_1(Au, u)$$

27

if and only if

$$C_0(A^{-1}u, u) \le (A^{-1}u, u) \le C_1(A^{-1}u, u),$$

where C_0 and C_1 are the same constants in both inequalities.

Proof: Using the Cauchy–Schwarz inequality we have

$$(A^{-1}u, u) = (AA^{-1}u, A^{-1}u) \le (AA^{-1}u, A^{-1}u)^{1/2}(A^{-1}u, u)^{1/2}$$

$$\le C_1^{1/2}(A^{-1}u, u)^{1/2}(A^{-1}u, u)^{1/2}.$$

Thus

$$(A^{-1}u, u) \le C_1(A^{-1}u, u).$$

The rest follows analogously.

Proof of Lemma 3.1: We first recall that for $\varphi, \psi \in M$

$$(A_k P_k \varphi, \psi) = (A_k P_k \varphi, Q_k \psi) = A(\varphi, Q_k \psi) = (A\varphi, Q_k \psi) = (Q_k A\varphi, \psi).$$

Hence $A_k P_k = Q_k A$.

Now suppose that A.1 is true for A. Then

$$(Av, v) \le C_a \left(A(P_1 v, v) + \sum_{k=2}^{J} \lambda_k^{-1} \|A_k P_k v\|^2 \right)$$

$$= C_a \left[(A_1^{-1} Q_1 Av, Q_1 Av) + \sum_{k=2}^{J} \lambda_k^{-1} \|Q_k Av\|^2 \right].$$

Now let $Av = w$. Then we have

(3.9)
$$(A^{-1}w, w) \le C_a \left[(A_1^{-1} Q_1 w, Q_1 w) + \sum_{k=2}^{J} \lambda_k^{-1} \|Q_k w\|^2 \right].$$

Let Λ_k be the largest eigenvalue of A_k. Then $\Lambda_k \le C_1 \lambda_k$ or $\lambda_k^{-1} \le C_1 \Lambda_k^{-1}$, and

(3.10)
$$C_0 \lambda_k \le \Lambda_k.$$

Also by Lemma 3.2 we have

$$(A_1^{-1}\psi, \psi) \le C_1(A_1^{-1}\psi, \psi), \text{ for all } \psi \in M_1$$

and

$$C_0(\mathcal{A}^{-1}w, w) \leq (A^{-1}w, w), \text{ for all } w \in M.$$

Hence

$$(\mathcal{A}^{-1}w, w) \leq \frac{C_1}{C_0} C_a \left[(\mathcal{A}_1^{-1}Q_1 w, Q_1 w) + \sum_{k=2}^{J} \Lambda_k^{-1} \|Q_k w\|^2 \right].$$

Setting $w = \mathcal{A}u$ implies that A.1 holds for $\mathcal{A}$ with the new $C_a \to \frac{C_1}{C_0} C_a$.

This lemma is important, especially in later applications.

So far we have imposed no conditions on the operators R_k. In order for them to be useful they must somehow be "smoothing" operators. A simple example of a smoothing operator is given in the two level algorithm of Section 2. The following is one such condition on R_k.

A.3: Let R_k be symmetric and, for constants a_0 and a_1, satisfy

$$a_0 \frac{\|u\|^2}{\lambda_k} \leq (R_k u, u) \leq a_1 \frac{\|u\|^2}{\lambda_k},$$

for all $u \in M_k$, for $k \geq 2$.

This class of smoothing operators allows for much greater flexibility than does the very simple looking smoother $\tilde{R}_k$. In particular it is not necessary to know λ_k exactly. In practical examples it is important to have such flexibility when considering questions of implementation. This is discussed in Section 10.

Instead of $\tilde{B}^a$ we define

$$(3.11) \qquad B^a = \sum_{k=1}^{J} R_k Q_k, \ R_1 = A_1^{-1}.$$

We prove now the simple result that B^a provides a uniform preconditioner for A.

THEOREM 3.1. *Let B^a be defined by (3.11). Suppose that A.1, A.2 and A.3 hold. Then*

$$K(B^a A) \leq \frac{a_1}{a_0} \left(\frac{2\tilde{C}}{1 - \epsilon} \right)^2 C_a.$$

Proof: A.1 implies that

$$A(u, u) \leq C_a \sum_{k=1}^{J} A(\tilde{T}_k u, u).$$

29

We shall show that, for some constant $\overline{C}_a$,

$$(3.12) \qquad \sum_{k=1}^{J} A(\tilde{T}_k u, u) \le \overline{C}_a A(u, u).$$

Recalling that

$$\tilde{T}_1 = P_1 \text{ and } \tilde{T}_k = \frac{1}{\lambda_k} A_k P_k \text{ for } k > 1$$

we have

$$\sum_{k=1}^{J} A(\tilde{T}_k u, u) = \sum_{k=1}^{J} \sum_{\ell=1}^{J} A(\tilde{T}_k u, (P_\ell - P_{\ell-1})u)$$

$$= \sum_{k=1}^{J} \sum_{\ell=1}^{k} A(\tilde{T}_k u, (P_\ell - P_{\ell-1})u)$$

$$\le \sum_{k=1}^{J} \sum_{\ell=1}^{k} A(\tilde{T}_k u, u)^{1/2} A(\tilde{T}_k w_\ell, w_\ell)^{1/2}$$

where $w_\ell = (P_\ell - P_{\ell-1})u \in M_\ell$, $\ell \le k$ and $P_0 = 0$. By A.2

$$A(\tilde{T}_k w_\ell, w_\ell) \le (\tilde{C}\epsilon^{k-\ell})^2 A(w_\ell, w_\ell).$$

Hence, setting $A(\tilde{T}_k u, u)^{1/2} = \alpha_k$ and $A(w_\ell, w_\ell)^{1/2} = \beta_\ell$,

$$\sum_{k=1}^{J} A(\tilde{T}_k u, u) \le \tilde{C} \sum_{k=1}^{J} \sum_{\ell=1}^{k} \epsilon^{k-\ell} \alpha_k \beta_\ell$$

$$\le \tilde{C} \sum_{k=1}^{J} \sum_{\ell=1}^{J} \epsilon^{|k-\ell|} \alpha_k \beta_\ell$$

$$= \tilde{C} < \mathcal{E}\vec{\alpha}, \vec{\beta} >.$$

Here $\mathcal{E}$ is the $J \times J$ matrix with positive entries $\epsilon^{|k-\ell|}$ and $< \cdot, \cdot >$ is the Euclidean inner product. Now the largest eigenvalue of $\mathcal{E}$ is bounded by its maximal row sum. That is

$$\lambda_{max}(\mathcal{E}) \le \frac{2}{1-\epsilon}.$$

Hence

$$< \mathcal{E}\vec{\alpha}, \vec{\beta} > \; \le \frac{2}{1-\epsilon} \Big(\sum_{k=1}^{J} \alpha_k^2\Big)^{1/2} \Big(\sum_{\ell=1}^{J} \beta_\ell^2\Big)^{1/2}.$$

Now we have that

$$\sum_{i=1}^{J} \alpha_i^2 \leq \frac{2\tilde{C}}{1-\varepsilon} \left(\sum_{k=1}^{J} \alpha_k^2 \right)^{1/2} \left(\sum_{\ell=1}^{J} \beta_\ell^2 \right)^{1/2}$$

or

$$\sum_{i=1}^{J} \alpha_i^2 \leq \left(\frac{2\tilde{C}}{1-\varepsilon} \right)^2 \sum_{i=1}^{J} \beta_i^2.$$

Also

$$\sum_{i=1}^{J} \beta_i^2 = \sum_{i=1}^{J} A((P_i - P_{i-1})u, (P_i - P_{i-1})u)$$

$$= \sum_{i=1}^{J} A((P_i - P_{i-1})u, u)$$

$$= A(u, u).$$

Hence we have shown that

$$C_a^{-1} A(u, u) \leq \sum_{k=1}^{J} A(\tilde{T}_k u, u) \leq \left(\frac{2\tilde{C}}{1-\varepsilon} \right)^2 A(u, u).$$

Assuming, without loss, that $a_0 \leq 1$ and $a_1 \geq 1$ in A.3, we see that for each k

$$\frac{1}{a_1} A(R_k A_k P_k u, u) \leq A(\tilde{T}_k u, u) \leq \frac{1}{a_0} A(R_k A_k P_k u, u)$$

and hence

(3.13) $$\qquad a_0 C_a^{-1} A(u, u) \leq A(B^a A u, u) \leq a_1 \left(\frac{2\tilde{C}}{1-\varepsilon} \right)^2 A(u, u).$$

This proves Theorem 3.1.

The following important corollary follows easily.

COROLLARY 3.1. *Let B^a be defined by (3.11). Suppose that A.1, A.2 and A.3 hold for some $\mathcal{A}$ equivalent to A. Then $K(B^a A)$ is bounded, with a bound which depends only on the constants in A.1, A.2 and A.3 and the constants in the equivalence relation between $\mathcal{A}$ and A.*

Proof: An inequality analogous to (3.13) holds for A replaced by $\mathcal{A}$. The corollary now follows by changing variables and using the equivalence between $\mathcal{A}$ and A.

Remark: Notice that the upper inequality in the additive preconditioner depends on the upper bound for R_k in A.3. For example, we might think of R_k as a preconditioner for A_k, e.g., $R_k = A_k^{-1}$. In this case, for $u \in M_k$

$$\|u\|^2 = (A_k A_k^{-1} u, u) \le \lambda_k(A_k^{-1} u, u) = \lambda_k(R_k u, u)$$

so that the lower inequality is satisfied. However the upper bound for $B^a A$ in this case is J. This is clear since

$$A(B^a A u, u) = \sum_{k=1}^{J} A(P_k u, u) = J A(u, u)$$

if $u \in M_1$ since $P_k u = u$ for all k. Thus the "smoother", in the additive case, must not be too good. This is in contrast to multiplicative algorithms such as the multigrid V–cycle.

We shall make somewhat weaker assumptions on R_k which will be useful in the multiplicative case. First, R_k need not be symmetric. It must, however, be properly scaled (we'll see what this means later).

Recall that $K_k = I - R_k A_k$ and $K_k^* = I - R_k^t A_k$. Note that for $k = 1$, $K_1 = K_1^* = 0$. The following conditions will be useful in proving estimates concerning the V–cycle algorithm.

A.4: There exists $C_R \ge 1$, independent of k such that

$$\frac{\|u\|^2}{\lambda_k} \le C_R(\overline{R}_k u, u), \text{ for all } u \in M_k,$$

where $\overline{R}_k = (I - K_k^* K_k) A_k^{-1}$.

First remark concerning A.4: Let $R_{k,\omega} = \omega \lambda_k^{-1} I$ and $K_{k,\omega} = I - R_{k,\omega} A_k$. Setting $\omega = \frac{1}{C_R}$ and $u = A_k v$, we have

$$A(R_{k,\omega} A_k v, v) \le A((I - K_k^* K_k) v, v).$$

Rewriting this, we have an alternative form of A.4.

32

A.4: There exists $\omega \in (0,1)$ such that

(3.14) $$A(K_k v, K_k v) \leq A(K_{k,\omega} v, v), \quad \text{for all } v \in M_k.$$

Clearly this implies that

$$A(K_k v, K_k v) \leq A(K_{k,\omega/2} v, K_{k,\omega/2} v).$$

This means that the smoothing process defined by $K_k = I - R_k A_k$ reduces the components of the error at least as fast as the simplest one for some $\tilde{\omega} = \omega/2 \in (0,1)$. Clearly (3.14) for some $\omega \in (0,1]$ implies A.4 with $C_R = \frac{1}{\omega}$.

The alternative form of A.4, (3.14), seems more natural since it says that the choice of smoother must be comparable to a simple smoother whose properties are transparent.

We also need the following scaling condition, for which we recall that $T_k = R_k A_k P_k$.

A.5: There exists a positive constant $\theta < 2$ such that

$$A(T_k v, T_k v) \leq \theta A(T_k v, v), \quad \text{for all } v \in M.$$

Remark concerning A.5: To see that such a condition is very natural we note, for example, that if $T_k = \frac{\alpha}{\lambda_k} A_k P_k$, then we need $0 < \alpha < 2$. If $u_k \in M_k$ is an eigenfunction corresponding to the largest eigenvalue λ_k of A_k, then $(I - T_k) u_k = (1 - \alpha) u_k$. Then if $\alpha \geq 2$, $|1 - \alpha| \geq 1$ and the operator $I - T_k$ will not reduce such a component of the error.

Second remark concerning A.4: It is easy to see that, with A.5, the lower inequality of A.3 alone implies A.4. Hence A.4 requires less of the smoother R_k and is thus a weaker condition.

The V–cycle estimates: The next part of this section (through Theorem 3.5) will be devoted to proving abstract results concerning the V–cycle algorithm. To this end recall, from the definition of B_J, that (3.4) holds. Set $E_0 = I$ and $E_i = (I - T_i)\ldots(I - T_1)$ for $i \geq 1$. Then

$$E_i = (I - T_i)E_{i-1}$$

and

$$E_J = (I - T_J)\ldots(I - T_1).$$

Thus (3.4) is the same as

$$I - B_J A_J = E_J E_J^*.$$

We shall prove first the following.

THEOREM 3.2. Let B_J be defined by Algorithm I. Assume A.1, A.2, A.4 and A.5. Then

$$0 \leq A((I - B_J A_J)v, v) \leq \left(1 - \frac{1}{C_m}\right) A(v, v)$$

where

$$C_m = 2C_a \left(C_R + \frac{\tilde{C}^2 \epsilon^2}{(1-\epsilon)^2} \frac{\theta}{2-\theta}\right).$$

Proof: The lower inequality is obvious since

$$A((I - B_J A_J)v, v) = A(E_J^* v, E_J^* v) \equiv |||E_J^* v|||^2 \geq 0.$$

Since $|||E_J||| = |||E_J^*|||$, we may estimate $|||E_J v|||$. Recall that $E_i = (I - T_i)E_{i-1}$. Hence

$$E_{i-1} - E_i = T_i E_{i-1}.$$

Summing from 1 to k, we have

(3.15)
$$I - E_k = \sum_{i=1}^{k} T_i E_{i-1}.$$

Clearly

$$A(E_{k-1}u, E_{k-1}u) = A(E_k u, E_k u) + 2A(E_k u, T_k E_{k-1}u) + A(T_k E_{k-1}u, T_k E_{k-1}u)$$

34

or

(3.16) $A(E_{k-1}u, E_{k-1}u) - A(E_k u, E_k u) = A((2I - T_k)E_{k-1}u, T_k E_{k-1}u).$

Let $\overline{T}_k = \overline{R}_k A_k P_k = (I - K_k^* K_k)P_k$. Thus, since $K_k = I - R_k A_k P_k$, we have

$$A(\overline{T}_k w, w) = A((I - (I - R_k^t A_k)(I - R_k A_k))P_k w, P_k w)$$
$$= ((R_k^t + R_k)A_k P_k w, A_k P_k w) - (R_k^t A_k R_k A_k P_k w, A_k P_k w)$$
$$= 2(R_k A_k P_k w, A_k P_k w) - (A_k R_k A_k P_k w, R_k A_k P_k w)$$
$$= A((2I - T_k)w, T_k w).$$

Hence taking $w = E_{k-1}v$

$$A(\overline{T}_k E_{k-1}v, E_{k-1}v) = A((2I - T_k)E_{k-1}v, T_k E_{k-1}v).$$

Summing (3.16) implies

(3.17) $A(v, v) - A(E_J v, E_J v) = \sum_{k=1}^{J} A(\overline{T}_k E_{k-1}v, E_{k-1}v).$

Note now that the upper estimate of Theorem 3.2 will follow if we can show that

$$A(v, v) \leq C_m[A(v, v) - A(E_J v, E_J v)].$$

We now use A.1, namely

$$A(v, v) \leq C_a \sum_{k=1}^{J} A(\tilde{T}_k v, v)$$

with $\tilde{T}_1 = P_1$ and $\tilde{T}_k = \frac{1}{\lambda_k} A_k P_k$ if $k > 1$. Hence

(3.18)

$$A(v, v) \leq 2C_a \left(\sum_{k=1}^{J} A(\tilde{T}_k E_{k-1}v, E_{k-1}v) + \sum_{k=2}^{J} A(\tilde{T}_k (I - E_{k-1})v, (I - E_{k-1})v) \right).$$

For $k = 2, \ldots, J$, A.4 implies that

$$A(\tilde{T}_k E_{k-1}v, E_{k-1}v) = \frac{\|A_k P_k E_{k-1}v\|^2}{\lambda_k}$$
$$\leq C_R(\overline{R}_k A_k P_k E_{k-1}v, A_k P_k E_{k-1}v)$$
$$= C_R A(\overline{R}_k A_k P_k E_{k-1}v, E_{k-1}v)$$
$$= C_R A(\overline{T}_k E_{k-1}v, E_{k-1}v).$$

35

Thus, since $\tilde{T}_1 = \overline{T}_1 = P_1$,

$$\sum_{k=1}^{J} A(\tilde{T}_k E_{k-1}v, E_{k-1}v) \leq C_R \sum_{k=1}^{J} A(\overline{T}_k E_{k-1}v, E_{k-1}v)$$
$$= C_R[A(v,v) - A(E_J v, E_J v)].$$

Hence for the first term on the right of (3.18) we have the desired estimate.

For the second term of (3.18) we use (3.15). That is

$$\sum_{k=2}^{J} A(\tilde{T}_k(I - E_{k-1})v, (I - E_{k-1})v)$$
$$= \sum_{k=2}^{J} \sum_{\ell=1}^{k-1} \sum_{m=1}^{k-1} A(\tilde{T}_k T_\ell E_{\ell-1}v, T_m E_{m-1}v)$$
$$\leq \sum_{k=2}^{J} \sum_{\ell=1}^{k-1} \sum_{m=1}^{k-1} A(\tilde{T}_k w_\ell, w_\ell)^{1/2} A(\tilde{T}_k w_m, w_m)^{1/2}$$

where we have used the Cauchy–Schwarz inequality and $w_\ell = T_\ell E_{\ell-1}v$. We continue the estimate using A.2 to obtain

$$\sum_{k=2}^{J} A(\tilde{T}_k(I - E_{k-1})v, (I - E_{k-1})v)$$
$$\leq \tilde{C}^2 \sum_{k=2}^{J} \sum_{\ell=1}^{k-1} \sum_{m=1}^{k-1} \varepsilon^{2k-\ell-m} A(w_\ell, w_\ell)^{1/2} A(w_m, w_m)^{1/2}.$$

Further, using the arithmetic–geometric mean inequality, we have

(3.19) $$\sum_{k=2}^{J} A(\tilde{T}_k(I - E_{k-1})v, (I - E_{k-1})v)$$
$$\leq \frac{\tilde{C}^2}{2} \sum_{k=2}^{J} \sum_{\ell=1}^{k-1} \sum_{m=1}^{k-1} \varepsilon^{2k-\ell-m} [A(w_\ell, w_\ell) + A(w_m, w_m)]$$
$$= \tilde{C}^2 \sum_{k=2}^{J} \sum_{\ell=1}^{k-1} \sum_{m=1}^{k-1} \varepsilon^{k-m} \varepsilon^{k-\ell} A(w_\ell, w_\ell)$$
$$\leq \frac{\varepsilon \tilde{C}^2}{1-\varepsilon} \sum_{k=2}^{J} \sum_{\ell=1}^{k-1} \varepsilon^{k-\ell} A(w_\ell, w_\ell).$$

36

Changing the order of summation and using A.5 it follows that

$$(3.20) \qquad \sum_{k=2}^{J} \sum_{\ell=1}^{k-1} \varepsilon^{k-\ell} A(w_\ell, w_\ell) = \sum_{\ell=1}^{J-1} \sum_{k=\ell+1}^{J} \varepsilon^{k-\ell} A(w_\ell, w_\ell)$$

$$\leq \frac{\varepsilon}{1-\varepsilon} \sum_{\ell=1}^{J-1} A(T_\ell E_{\ell-1} v, T_\ell E_{\ell-1} v)$$

$$\leq \frac{\theta \varepsilon}{1-\varepsilon} \sum_{\ell=1}^{J-1} A(T_\ell E_{\ell-1} v, E_{\ell-1} v).$$

Now $\overline{T}_k = \overline{R}_k A_k P_k$ and by A.5

$$(3.21) \qquad A(\overline{T}_k w, w) = A((2I - T_k)w, T_k w) \geq (2 - \theta) A(T_k w, w).$$

Hence, combining (3.19), (3.20) and (3.21), we see that

$$\sum_{k=2}^{J} A(\tilde{T}_k (I - E_{k-1}) v, (I - E_{k-1}) v) \leq \frac{\theta}{2-\theta} \left(\frac{\varepsilon}{1-\varepsilon}\right)^2 \tilde{C}^2 \sum_{k=1}^{J-1} A(\overline{T}_k E_{k-1} v, E_{k-1} v)$$

and, since $A(\overline{T}_J \varphi, \varphi) = (\overline{R}_J A\varphi, A\varphi) \geq 0$ for any φ, we conclude that

$$\sum_{k=2}^{J} A(\tilde{T}_k (I - E_{k-1}) v, (I - E_{k-1}) v) \leq \frac{\theta}{2-\theta} \left(\frac{\varepsilon}{1-\varepsilon}\right)^2 \tilde{C}^2 \sum_{k=1}^{J} A(\overline{T}_k E_{k-1} v, E_{k-1} v).$$

Thus we have shown that

$$A(v, v) \leq C_m \sum_{k=1}^{J} A(\overline{T}_k E_{k-1} v, E_{k-1} v)$$

$$= C_m [A(v, v) - A(E_J v, E_J v)].$$

Hence

$$A(E_J v, E_J v) \leq (1 - 1/C_m) A(v, v),$$

from which it follows that

$$A((I - A_J B_J) v, v) \leq (1 - 1/C_m) A(v, v),$$

where

$$C_m = 2C_a \left[C_R + \frac{\theta}{2-\theta} \left(\frac{\varepsilon}{1-\varepsilon}\right)^2 \tilde{C}^2 \right].$$

Smoothing on subspaces: In some applications such as are found in Section 6 it is important to be able to refine only certain portions of the mesh. In such cases it is convenient to be able to use multigrid algorithms in which smoothing is performed only on the refined portion of the mesh. We shall consider this here. The V–cycle multigrid algorithm already allows for such a situation by properly defining R_k. We shall formalize this as follows.

As always, so far, we have

$$M_1 \subset M_2 \subset \cdots \subset M_J = M.$$

Associated with each M_i consider a subspace $\hat{M}_i \subseteq M_i$ with $\hat{M}_1 = M_1$. The $\hat{M}_i$'s are not necessarily nested and they are to be thought of as the places where the smoothing will be done. For the purpose of formulating relevant conditions we define $\hat{A}_k : \hat{M}_k \to \hat{M}_k, \hat{P}_k : M \to \hat{M}_k$ and $\hat{Q}_k : M \to \hat{M}_k$ by

$$(\hat{A}_k \varphi, \chi) = A(\varphi, \chi), \text{ for all } \chi \in \hat{M}_k,$$

$$A(\hat{P}_k v, \chi) = A(v, \chi), \text{ for all } \chi \in \hat{M}_k$$

and

$$(\hat{Q}_k v, \chi) = (v, \chi), \text{ for all } \chi \in \hat{M}_k.$$

Define further $\hat{T}_k = \frac{1}{\lambda_k} \hat{A}_k \hat{P}_k = \frac{1}{\lambda_k} \hat{Q}_k A$, $\hat{T}_1 = P_1$. Now we have the more general conditions.

A.1a:
$$A(u, u) \leq C_a \left[\sum_{k=1}^{J} A(\hat{T}_k u, u) \right], \text{ for all } u \in M.$$

A.2a:
$$A(\hat{T}_k w, w) \leq (\hat{C} \varepsilon^{k-\ell})^2 A(w, w),$$

for all $w \in \hat{M}_\ell, \ell \leq k.$

Smoothing on $\hat{M}_k$: Now suppose that $R_k : M_k \to \hat{M}_k.$

38

A.3a: Assume that R_k is symmetric and there are constants a_0 and a_1 such that

$$a_0 \frac{\|v\|^2}{\lambda_k} \le (R_k v, v) \le a_1 \frac{\|v\|^2}{\lambda_k}, \text{ for all } v \in \hat{M}_k.$$

Note that in this case

$$(R_k \varphi, \psi) = (\varphi, R_k \psi) = (\hat{Q}_k \varphi, R_k \psi) = (R_k \hat{Q}_k \varphi, \psi),$$

for all $\varphi, \psi \in M_k$. Hence

$$R_k = R_k \hat{Q}_k.$$

The generalization of A.4 is the following.

A.4a: Assume that $R_k = R_k \hat{Q}_k$ and

$$\frac{\|u\|^2}{\lambda_k} \le C_R (\overline{R}_k u, u), \text{ for all } u \in \hat{M}_k$$

with $\overline{R}_k = (I - K_k^* K_k) A_k^{-1}$.

Note that on M_k

$$K_k = I - R_k A_k = I - R_k \hat{Q}_k A_k = I - R_k \hat{A}_k \hat{P}_k.$$

Now the multigrid V–cycle algorithm is the same as before. The additive preconditioner is the same if we define

$$T_k = R_k A_k P_k = R_k \hat{A}_k \hat{P}_k.$$

We state the results in the following:

THEOREM 3.1s. *Let B^a be defined by (3.11). Suppose that A.1a, A.2a and A.3a hold for some A equivalent to A. Then*

$$K(B^a A) \le \frac{a_1}{a_0} \left(\frac{2\hat{C}}{1-\varepsilon} \right)^2 C_a.$$

THEOREM 3.2s. *Let B_J be defined by (3.11). Assume A.1a, A.2a, A.4a and A.5. Then*

$$0 \le A((I - B_J A_J)v, v) \le \left(1 - \frac{1}{C_m} \right) A(v, v)$$

39

where

$$C_m = 2C_a\Big(C_R + \frac{\hat{C}^2\varepsilon^2}{(1-\varepsilon)^2}\frac{\theta}{2-\theta}\Big).$$

The proofs of Theorems 3.1s and 3.2s are virtually the same as the proofs of Theorems 3.1 and 3.2. We will see that this generalization yields interesting results for some examples in Section 6.

The previous results all involved the hypotheses that either A.1 or A.2 (or their generizations A.1a or A.2a) were satisfied. Though these are not very restrictive, there are cases in which, for example, A.1 or A.2 do not appear to hold. For example, in the application to second order elliptic equations, when the coefficients have large jumps, A.1 is not known to hold independently of the size of the jumps. In the case in which the coefficients are not very smooth, e.g. they are only bounded and measurable, it is not known whether A.2 holds. A.1 holds in such a case by virtue of Lemma 3.1. In the application of Section 8 to pseudodifferential operators of order minus one, neither A.1 nor A.2 is known to hold in the general case. Hence we shall next consider a very weak hypothesis which yields some results for the above mentioned cases. Because of the weakness of the new hypothesis the results are also somewhat weaker. In fact, the contraction number for the V–cycle algorithm may deteriorate linearly in the number of levels.

We now state our weak hypothesis.

A.6: Assume that there exist linear operators $\overline{Q}_k : M \to M_k$, $k = 1,\ldots,J$ with $\overline{Q}_J = I$, such that

$$\|(\overline{Q}_k - \overline{Q}_{k-1})u\|^2 \le C\lambda_k^{-1}A(u,u),\ \ k = 2,\ldots,J$$

and

$$A(\overline{Q}_k u, \overline{Q}_k u) \le C A(u,u),\ \ k = 1,\ldots,J.$$

<u>Remark:</u>The operators $\overline{Q}_k$, in some applications may be the operators Q_k themselves. In the refinement application, however, more care must be taken in their choice.

With A.6 we may prove the following.

THEOREM 3.3. *Let B_J be defined by Algorithm I. Suppose that A.4a, A.5 and A.6 are satisfied and that $\hat{M}_k \supseteq \text{Range}(\overline{Q}_k - \overline{Q}_{k-1})$. Then*

$$A((I - B_J A_J)v, v) \le \left(1 - \frac{1}{CJ}\right) A(v, v)$$

for some constant C independent of J.

Proof:

$$(3.22) \qquad A(u, u) = \sum_{k=2}^{J} A(u, (\overline{Q}_k - \overline{Q}_{k-1})u) + A(u, \overline{Q}_1 u)$$

$$= \sum_{k=2}^{J} A(E_{k-1}u, (\overline{Q}_k - \overline{Q}_{k-1})u) + A(u, \overline{Q}_1 u)$$

$$+ \sum_{k=2}^{J} A((I - E_{k-1})u, (\overline{Q}_k - \overline{Q}_{k-1})u).$$

First we have

$$(3.23)$$

$$\sum_{k=2}^{J} A(E_{k-1}u, (\overline{Q}_k - \overline{Q}_{k-1})u) = \sum_{k=2}^{J} (\hat{A}_k \hat{P}_k E_{k-1}u, (\overline{Q}_k - \overline{Q}_{k-1})u)$$

$$\le \sum_{k=2}^{J} \|\hat{A}_k \hat{P}_k E_{k-1}u\| \, \|(\overline{Q}_k - \overline{Q}_{k-1})u\|$$

$$\le C \left(\sum_{k=2}^{J} \lambda_k \|(\overline{Q}_k - \overline{Q}_{k-1})u\|^2\right)^{1/2} \left(\sum_{k=2}^{J} \lambda_k^{-1} \|\hat{A}_k \hat{P}_k E_{k-1}u\|^2\right)^{1/2}.$$

Using A.4a, A.6 and the fact that $\overline{T}_k = \overline{R}_k A_k P_k = \overline{R}_k \hat{A}_k \hat{P}_k$, we see that

$$\sum_{k=2}^{J} A(E_{k-1}u, (\overline{Q}_k - \overline{Q}_{k-1})u) \le CJ^{1/2} A^{1/2}(u, u) \left[\sum_{k=1}^{J} A(\overline{T}_k E_{k-1}u, E_{k-1}u)\right]^{1/2}.$$

Now we estimate the remaining terms

$$\sum_{k=2}^{J} A((I - E_{k-1})u, (\overline{Q}_k - \overline{Q}_{k-1})u) + A(u, \overline{Q}_1 u).$$

41

Recall that $I - E_k = \sum_{\ell=1}^{k} T_\ell E_{\ell-1}$. Rearranging the sum, we have

$$\sum_{k=2}^{J} A((I - E_{k-1})u, (\overline{Q}_k - \overline{Q}_{k-1})u) + A(u, \overline{Q}_1 u)$$

$$= \sum_{k=2}^{J} A((I - E_{k-1})u, \overline{Q}_k u) - \sum_{k=1}^{J-1} A((I - E_k)u, \overline{Q}_k u) + A(u, \overline{Q}_1 u)$$

$$= \sum_{k=2}^{J-1} A((E_k - E_{k-1})u, \overline{Q}_k u) + A((I - E_{J-1})u, u).$$

Now $E_k - E_{k-1} = -T_k E_{k-1}$ and hence

$$\sum_{k=2}^{J} A((I - E_{k-1})u, (\overline{Q}_k - \overline{Q}_{k-1})u) + A(u, \overline{Q}_1 u)$$

$$= -\sum_{k=2}^{J-1} A(T_k E_{k-1}u, \overline{Q}_k u) + \sum_{k=1}^{J-1} A(T_k E_{k-1}u, u)$$

$$= \sum_{k=2}^{J-1} A(T_k E_{k-1}u, (P_k - \overline{Q}_k)u) + A(T_1 u, u).$$

Hence, using the Cauchy–Schwarz inequality,

$$(3.24) \qquad \sum_{k=2}^{J} A((I - E_{k-1})u, (\overline{Q}_k - \overline{Q}_{k-1})u) + A(u, \overline{Q}_1 u)$$

$$\leq \left(\sum_{k=1}^{J} A(T_k E_{k-1}u, T_k E_{k-1}u) \right)^{1/2}$$

$$\cdot \left(\sum_{k=2}^{J} A((P_k - \overline{Q}_k)u, (P_k - \overline{Q}_k)u) + A(u, u) \right)^{1/2}$$

$$\leq C J^{1/2} A^{1/2}(u, u) \left(\sum_{k=1}^{J} A(\overline{T}_k E_{k-1}u, E_{k-1}u) \right)^{1/2}.$$

Here we used A.5, A.6 and the boundedness of P_k in $A(\cdot, \cdot)$. Hence

$$A(u, u) \leq C J \left[\sum_{k=1}^{J} A(\overline{T}_k E_{k-1}u, E_{k-1}u) \right] = C J [A(u, u) - A(E_J u, E_J u)].$$

This implies that

$$A(E_J u, E_J u) \leq \left(1 - \frac{1}{CJ} \right) A(u, u)$$

and thus proves Theorem 3.3.

Looking at the proof of Theorem 3.3, we see that the dependence of the estimate on J comes from the bounds for $\sum_{k=2}^{J} \lambda_k \|(\overline{Q}_k - \overline{Q}_{k-1})u\|^2$ in (3.23) and $\sum_{k=2}^{J} A((P_k - \overline{Q}_k)u, (P_k - \overline{Q}_k)u)$ in (3.24). We shall consider another condition which can be used instead of A.1 and A.2 and which can be used to obtain uniform bounds for these quantities. Such bounds lead immediately to a uniform contraction result.

In order to formulate the next condition we define, as usual, the powers of a positive definite operator via the corresponding spectral representation. That is, for $\{\Lambda_i\}$ and $\{\varphi_i\}$ as in Section 1, we have

$$u = \sum_{i=1}^{N} (u, \varphi_i)\varphi_i,$$

$$Au = \sum_{i=1}^{N} \Lambda_i (u, \varphi_i)\varphi_i$$

and

$$A^\alpha u = \sum_{i=1}^{N} \Lambda_i^\alpha (u, \varphi_i)\varphi_i.$$

Our new condition is as follows.

A.7: For some $\alpha \in (0, 1]$ there is a constant C_α independent of k such that

$$(A_k^{1-\alpha} Q_k u, Q_k u) \le C_\alpha (A^{1-\alpha} u, u)$$

and

$$(A^{1-\alpha}(I - P_k)u, (I - P_k)u) \le C_\alpha \lambda_k^{-\alpha} A(u, u),$$

for all $u \in M$.

Remark: In the case of the model problem of Section 2, with Ω a convex polygonal domain, A.7 holds with $\alpha = 1$. Note that in this case the first inequality is trivial. If the boundary is not convex, The second inequality holds for some $\alpha < 1$ and the first inequality may be proved by interpolation. First prove that

$$(A_k Q_k u, Q_k u) \le C_1^2 (Au, u)$$

43

and note the trivial inequality

$$\|Q_k u\|_0^2 \leq \|u\|_0^2.$$

These inequalities say that $Q_k : M \to M_k$ is bounded as a map from M (with norm $(A \cdot, \cdot)^{1/2}$) to M_k (with norm $(A_k \cdot, \cdot)^{1/2}$) and bound C_1. Also it is bounded in $\| \cdot \|_0$. Hence, by Theorem B.4 on interpolation of operators,

$$(A_k^{1-\alpha} Q_k u, Q_k u) \leq (C_1^2)^{1-\alpha} (A^{1-\alpha} u, u).$$

The other inequality of A.7 may be proved via the Aubin–Nitsche duality argument. We shall show how A.7 may be verified in some examples in Sections 6, 7 and 8.

We now prove a result based on the hypothesis A.7.

THEOREM 3.4. *Let B_J be defined by Algorithm I. Suppose that A.7, A.4 and A.5 hold. Suppose further that there are positive constants $\delta < 1$ and $\tilde{C}$ such that $\lambda_k / \lambda_\ell \leq \tilde{C} \delta^{\ell-k}$, for all ℓ and k with $\ell \geq k - 1, k \geq 2$. Then there is a constant $C > 1$ and independent of J such that*

$$A((I - B_J A_J) u, u) \leq \left(1 - \frac{1}{C}\right) A(u, u).$$

Proof: Examining the proof of Theorem 3.3 we see that we can take $\hat{M}_k = M_k$ and $\overline{Q}_k = Q_k$. As mentioned above, the theorem will follow if we obtain the estimates

(3.25) $$\sum_{k=2}^{J} \lambda_k \|(Q_k - Q_{k-1}) u\|^2 \leq C A(u, u)$$

and

(3.26) $$\sum_{k=2}^{J} \|\|(P_k - Q_k) u\|\|^2 \leq C A(u, u).$$

We will show that these estimates follow from A.7.

44

We first prove (3.26). Clearly

$$\sum_{k=1}^{J} |||(P_k - Q_k)u|||^2 = \sum_{k=1}^{J} |||Q_k(I - P_k)u|||^2$$

$$\leq \sum_{k=1}^{J} \lambda_k^\alpha (A_k^{1-\alpha} Q_k(I - P_k)u, Q_k(I - P_k)u)$$

$$\leq C_\alpha \sum_{k=1}^{J} \lambda_k^\alpha (A^{1-\alpha}(I - P_k)u, (I - P_k)u)$$

$$= C_\alpha \sum_{k=1}^{J} \sum_{\ell=k+1}^{J} \sum_{m=k+1}^{J} \lambda_k^\alpha (A^{1-\alpha}(P_\ell - P_{\ell-1})u, (P_m - P_{m-1})u)$$

where we used the first part of A.7 and the identity $I - P_k = \sum_{\ell=k+1}^{J}(P_\ell - P_{\ell-1})$.
Applying the Cauchy–Schwarz inequality to each term of the sum yields

$$\sum_{k=1}^{J} |||(P_k - Q_k)u|||^2 \leq C_\alpha \sum_{k=1}^{J} \sum_{\ell,m=k+1}^{J} \lambda_k^\alpha \big((A^{1-\alpha}(P_\ell - P_{\ell-1})u, (P_\ell - P_{\ell-1})u)^{1/2}$$

$$\cdot (A^{1-\alpha}(P_m - P_{m-1})u, (P_m - P_{m-1})u)^{1/2}\big).$$

By the second inequality of A.7 we have, since $P_\ell - P_{\ell-1} = (I - P_{\ell-1})(P_\ell - P_{\ell-1})$,
we have

$$\sum_{k=1}^{J} |||(P_k - Q_k)u|||^2$$

$$\leq C_\alpha^2 \sum_{k=1}^{J} \sum_{\ell,m=k+1}^{J} \left(\frac{\lambda_k^2}{\lambda_{\ell-1}\lambda_{m-1}}\right)^{\alpha/2} |||(P_\ell - P_{\ell-1})u||| \, |||(P_m - P_{m-1})u|||.$$

Now since ℓ and m are greater than k, by hypothesis $\lambda_k/\lambda_{\ell-1} \leq \tilde{C}\delta^{\ell-1-k}$ and
$\lambda_k/\lambda_{m-1} \leq \tilde{C}\delta^{m-1-k}$. Hence

$$\sum_{k=1}^{J} |||(P_k - Q_k)u|||^2$$

$$\leq \frac{C_\alpha^2 \tilde{C}^\alpha \delta^{-\alpha}}{2} \sum_{k=1}^{J} \sum_{\ell,m=k+1}^{J} \delta^{(\ell+m-2k)\alpha/2}(|||(P_\ell - P_{\ell-1})u|||^2 + |||(P_m - P_{m-1})u|||^2).$$

45

Since the two resulting sums are the same, we have

$$\sum_{k=1}^{J} |||(P_k - Q_k)u|||^2 \le C_\alpha^2 \tilde{C}^\alpha \delta^{-\alpha} \sum_{k=1}^{J} \sum_{\ell=k+1}^{J} \sum_{m=k+1}^{J} \delta^{(\ell+m-2k)\alpha/2} |||(P_\ell - P_{\ell-1})u|||^2.$$

Summing over m and interchanging the order, we have

$$\sum_{k=1}^{J} |||(P_k - Q_k)u|||^2$$

$$\le C_\alpha^2 \tilde{C}^\alpha \delta^{-\alpha} \Big(\sum_{j=1}^{\infty} \delta^{j\alpha/2} \Big) \Big(\sum_{\ell=2}^{J} |||(P_\ell - P_{\ell-1})u|||^2 \sum_{k=1}^{\ell-1} \delta^{(\ell-k)\alpha/2} \Big)$$

$$\le C A(u, u)$$

with $C = C_\alpha^2 \tilde{C}^\alpha (1 - \delta^{\alpha/2})^{-2}$. This proves (3.26).

We complete the proof of the theorem by proving (3.25). By A.7

$$\inf_{\chi \in M_k} (A^{1-\alpha}(v - \chi), (v - \chi)) \le (A^{1-\alpha}(I - P_k)v, (I - P_k)v) \le C_\alpha \lambda_k^{-\alpha} A(v, v)$$

which implies, using the simultaneous approximation result, Theorem B.5, that

$$\|(I - Q_k)v\|^2 = \inf_{\chi \in M_k} \|v - \chi\|^2 \le C \lambda_k^{-1} A(v, v).$$

Thus, since $Q_k - Q_{k-1} = (I - Q_{k-1})(Q_k - Q_{k-1})$, we see that

$$\lambda_k^{1/2} \|(Q_k - Q_{k-1})u\|$$

$$\le (\tilde{C}/\delta)^{1/2} \lambda_{k-1}^{1/2} \|(Q_k - Q_{k-1})u\| \le C |||(Q_k - Q_{k-1})u|||$$

$$\le C(|||(P_k - Q_k)u||| + |||(P_{k-1} - Q_{k-1})u||| + |||(P_k - P_{k-1})u|||).$$

Squaring and summing and using (3.26) yields (3.25).

Theorem 3.4 extends easily to the case in which we smooth only on subspaces as follows. In the proof of Theorem 3.3 we wrote

$$A(u, u) = \sum_{k=1}^{J} A(u, (\overline{Q}_k - \overline{Q}_{k-1})u)$$

$$= \sum_{k=1}^{J} (\hat{A}_k \hat{P}_k E_{k-1} u, (\overline{Q}_k - \overline{Q}_{k-1})u) + \sum_{k=2}^{J-1} A(T_k E_{k-1} u, (I - \overline{Q}_k)u)$$

$$+ A(T_1 u, u)$$

46

where $\mathrm{Range}(\overline{Q}_k - \overline{Q}_{k-1}) \subseteq \hat{M}_k$. Now suppose that we have

$$\hat{M}_k \subseteq M_k \subseteq \overline{\overline{M}}_k, \ \overline{\overline{M}}_k \subseteq \overline{\overline{M}}_{k+1}.$$

Here we want to think of the spaces M_k as the locally refined spaces, the spaces $\hat{M}_k$ as the refinement portion and the spaces $\overline{\overline{M}}_k$ as spaces fully refined which are introduced purely for theoretical purposes but do not enter into the algorithm.

We now state the assumptions analogous to A.7 but allowing local refinements.

A.7a: Assume that there exist linear operators $\overline{Q}_k : M \to M_k$, $k = 1, \ldots, J$ with $\overline{Q}_J = I$ and $\hat{M}_k \supseteq \mathrm{Range}(\overline{Q}_k - \overline{Q}_{k-1})$.

Assume further that for some $\alpha \in (0,1]$

$$(\overline{\overline{A}}^{1-\alpha}(I - \overline{\overline{P}}_k)v, (I - \overline{\overline{P}}_k)v) \leq C\lambda_k^{-\alpha} A(v,v), \ \text{for all } v \in M \subset \overline{\overline{M}}$$

and

$$(\overline{\overline{A}}_k^{1-\alpha}\overline{Q}_k v, \overline{Q}_k v) \leq C(\overline{\overline{A}}^{1-\alpha}v, v), \ \text{for all } v \in M,$$

where λ_k is the largest eigenvalue of $\overline{\overline{A}}_k$, and $\overline{\overline{A}}_k$, $\overline{\overline{P}}_k$ and $\overline{Q}_k$ are defined analogously.

Another technical condition which we will need and which is satisfied in our refinement example of Section 6 is the following.

A.8:

$$\|(I - \overline{Q}_k)u\| \leq C\|(I - \overline{\overline{Q}}_k)u\|,$$

for all $u \in M$. Then

THEOREM 3.5. *Let B_J be defined by Algorithm I. Suppose that A.4a, A.5, A.7a and A.8 hold. Then there is a constant $C > 1$ and independent of J such that*

$$A((I - B_J A_J)u, u) \leq \left(1 - \frac{1}{C}\right)A(u,u),$$

for all $u \in M$.

Proof: We need to show that

$$\sum_{k=1}^{J} \lambda_k \|(\overline{Q}_k - \overline{Q}_{k-1})u\|^2 \leq CA(u,u)$$

47

and

$$\sum_{k=2}^{J} |||\overline{\overline{P}}_k - \overline{Q}_k)u|||^2 \le CA(u,u).$$

The first of these inequalities follows from

$$\sum_{k=1}^{J} \lambda_k \|(\overline{Q}_k - \overline{Q}_{k-1})u\|^2 \le C \sum_{k=1}^{J} \lambda_k \|(I - \overline{\overline{Q}}_k)u\|^2 \le CA(u,u)$$

which, in turn, follows from the proof of Theorem 3.4. For the second inequality we have that

$$\sum_{k=2}^{J-1} A((\overline{\overline{P}}_k - \overline{Q}_k)u, (\overline{\overline{P}}_k - \overline{Q}_k)u)$$

$$\le C\Big(\sum_{k=2}^{J-1} A((\overline{\overline{P}}_k - \overline{\overline{Q}}_k)u, (\overline{\overline{P}}_k - \overline{\overline{Q}}_k)u) + \sum_{k=2}^{J} \lambda_k \|(\overline{\overline{Q}}_k - \overline{Q}_k)u\|^2\Big)$$

$$\le C\Big(A(u,u) + \sum_{k=1}^{J-1} \lambda_k \|(I - \overline{\overline{Q}}_k)u\|^2\Big)$$

$$\le CA(u,u),$$

by the argument in the proof of Theorem 3.4. This completes the proof of Theorem 3.5.

We have introduced several different conditions which were used as hypotheses to prove certain estimates for the additive preconditioner B^a and the multigrid V-cycle. We saw that A.7 implies (3.25). It is interesting to compare (3.25) with A.1. We will now examine the relationship between these inequalities.

To do this let

$$B = \sum_{\ell=1}^{J} \lambda_\ell^{-1}(Q_\ell - Q_{\ell-1}), \quad Q_0 = 0.$$

Then

$$B^{-1} = \sum_{k=1}^{J} \lambda_k(Q_k - Q_{k-1}).$$

This is clear since

$$(Q_\ell - Q_{\ell-1})(Q_k - Q_{k-1}) = \begin{cases} 0, & \text{if } k \ne \ell \\ Q_k - Q_{k-1}, & \text{if } k = \ell, \end{cases}$$

48

since $Q_\ell Q_k = Q_k Q_\ell = Q_\ell$ if $\ell \le k$. Hence (3.25) is $(B^{-1}u, u) \le C_1(Au, u)$ and, by Lemma 3.2, this is equivalent to

$$(A^{-1}u, u) \le C_1(Bu, u),$$

or, changing variables,

$$(Au, u) \le C_1 A(BAu, u).$$

Consider

$$\hat{B} = \sum_{k=1}^{J} \lambda_k^{-1} Q_k.$$

Write

$$B = \sum_{\ell=1}^{J-1} (\lambda_\ell^{-1} - \lambda_{\ell+1}^{-1}) Q_\ell + \lambda_J^{-1}.$$

Then clearly $(Bu, u) \le (\hat{B}u, u)$ and, assuming that $\lambda_k \le \delta_1 \lambda_{k+1}$, for some $\delta_1 < 1$, it follows that

$$(1 - \delta_1) \frac{1}{\lambda_k} \le \frac{1}{\lambda_k} - \frac{1}{\lambda_{k+1}}.$$

Thus

$$(1 - \delta_1)(\hat{B}u, u) \le (Bu, u).$$

Consequently

$$(Au, u) \le C_1 A(\hat{B}Au, u)$$
$$= C_1 \left[\sum_{k=2}^{J} \lambda_k^{-1} \|A_k P_k u\|^2 + \lambda_1^{-1} \|A_1 P_1 u\|^2 \right]$$
$$\le C_1 \left[\sum_{k=2}^{J} A(\tilde{T}_k u, u) + A(P_1 u, u) \right] = C_1 \sum_{k=1}^{J} A(\tilde{T}_k u, u).$$

Hence (3.25) implies A.1. Clearly

$$A(P_1 u, u) = (A^{-1} A_1 P_1 u, A_1 P_1 u) \le \left(\frac{\lambda_1}{\lambda_{min}} \right) \frac{\|A_1 P_1 u\|^2}{\lambda_1}.$$

So if $\frac{\lambda_1}{\lambda_{min}} \leq C$ then A.1 implies that

$$A(u, u) \leq C \left[\sum_{k=2}^{J} A(\tilde{T}_k u, u) + \frac{\|A_1 P_1 u\|^2}{\lambda_1} \right]$$

$$= C \sum_{k=1}^{J} \lambda_k^{-1} \|A_k P_k u\|^2 = C \sum_{k=1}^{J} \lambda_k^{-1} \|Q_k A u\|^2.$$

Therefore

$$(A^{-1} v, v) \leq C \sum_{k=1}^{J} \lambda_k^{-1} \|Q_k v\|^2 = C(\hat{B} v, v) \leq \frac{C}{1 - \delta_1} (B v, v).$$

Inequality (3.25) follows using Lemma 3.2, i.e.,

$$(B^{-1} v, v) \leq \frac{C}{1 - \delta_1} (A v, v).$$

Thus, with a mild growth condition on how λ_k increases with k, we find that A.1 is equivalent to (3.25).

So far we have looked only at the V–cycle with nested spaces and one smoothing, i.e. Algorithm I. We next generalize this algorithm, allowing more than one smoothing step and more than one correction step.

Other multigrid algorithms:

As before we will define an operator $B = B_J$. We will do this by induction. We shall still consider the case $M_1 \subset \ldots \subset M_J = M$.

Algorithm II: Let m and p be positive integers.

0) $B_1 = A_1^{-1}$ and suppose $R_k = R_k^t$.

For $g \in M_k$ we define $B_k g$ in terms of B_{k-1}, $2 \leq k \leq J$.

1) Set $x^0 = 0$, $q^0 = 0$

2) Define x^ℓ for $\ell = 1, \ldots, m$ by

$$x^\ell = x^{\ell-1} + R_k(g - A_k x^{\ell-1})$$

(smooth m times)

3) Define $y^m = x^m + q^p$, where q^i for $i = 1, \ldots, p$ is defined by

$$q^i = q^{i-1} + B_{k-1}[Q_{k-1}(g - A_k x^m) - A_{k-1} q^{i-1}]$$

(correct p times)

4) Define y^ℓ for $\ell = m+1, \ldots, 2m$ by

$$y^\ell = y^{\ell-1} + R_k(g - A_k y^{\ell-1})$$

(smooth m times)

5) Set $B_k g = y^{2m}$.

Remark: The case $p = 2$ is the so–called W–cycle. It was one of the earliest multigrid algorithms studied.

As in the V–cycle, Algorithm I, we want to study the operator $I - B_k A_k$. We shall obtain a recurrence relation analogous to (3.3) and then show that $I - B_J A_J$ is again a product of operators each of which is defined on all of M. To this end let $A_k x = g$ and $K_k = I - R_k A_k$. Then

$$x - x^\ell = K_k(x - x^{\ell-1}), \quad x^0 = 0$$

and hence

$$x - x^m = K_k^m x.$$

Now

$$x - y^m = x - x^m - q^p,$$

where

$$q^i = q^{i-1} + B_{k-1} A_{k-1} P_{k-1}[(x - x^m) - q^{i-1}]$$

or

$$x - x^m - q^i = (I - B_{k-1} A_{k-1} P_{k-1})[x - x^m - q^{i-1}].$$

Thus

$$x - y^m = x - x^m - q^p = (I - B_{k-1} A_{k-1} P_{k-1})^p K_k^m x.$$

The fourth step yields

$$x - y^\ell = K_k(x - y^{\ell-1}), \quad \ell = m+1, \dots, 2m$$

and

$$x - y^{2m} = K_k^m(x - y^m).$$

Hence

$$I - B_k A_k = K_k^m(I - B_{k-1}A_{k-1}P_{k-1})^p K_k^m.$$

Now, as before,

$$I - B_{k-1}A_{k-1}P_{k-1} = I - P_{k-1} + (I - B_{k-1}A_{k-1})P_{k-1}$$

and since $(I - P_{k-1})v = 0$ for all $v \in M_{k-1}$,

$$(I - B_{k-1}A_{k-1}P_{k-1})^p = (I - P_{k-1}) + (I - B_{k-1}A_{k-1})^p P_{k-1}.$$

Thus

(3.27) $\qquad I - B_k A_k = K_k^m[(I - P_{k-1}) + (I - B_{k-1}A_{k-1})^p P_{k-1}]K_k^m.$

We now want to see that the previous theorems are valid for $p \geq 1$ and $m \geq 1$ at least for $K_k = K_k^*$. As before, we shall see that $I - B_J A_J$ is a product. Extend K_k by $I - R_k A_k P_k$ as before. Then

$$I - B_k A_k P_k = I - P_k + K_k^m(I - B_{k-1}A_{k-1}P_{k-1})^p K_k^m P_k$$
$$= I - P_k + K_k^m(P_k - B_{k-1}A_{k-1}P_{k-1})^p K_k^m.$$

Now

$$(I - P_k)K_k^m = I - P_k$$

and hence

$$K_k^m(I - P_k)K_k^m = I - P_k.$$

Hence

$$I - B_k A_k P_k = K_k^m [(I - P_k) + (P_k - B_{k-1} A_{k-1} P_{k-1})^p] K_k^m$$

$$= K_k^m (I - B_{k-1} A_{k-1} P_{k-1})^p K_k^m.$$

Setting $T_k = I - K_k^m$ and $F_k = I - B_k A_k P_k$, we have

$$F_k = (I - T_k) F_{k-1}^p (I - T_k).$$

Thus

$$F_k = (I - T_k)(I - T_{k-1}) F_{k-2}^p (I - T_{k-1}) F_{k-1}^{p-1} (I - T_k).$$

Assume that $\sigma(K_k) \subset [0,1]$. Hence by induction F_k is symmetric in $A(\cdot, \cdot)$ and $\sigma(F_k) \subset [0,1]$. Thus we see that $F_J = (I - T_J) \cdots (I - T_1) \tilde{E}$ where $\|\tilde{E}\|_A \le 1$. Therefore any contraction result for $E_J = (I - T_J) \cdots (I - T_1)$ implies a result for F_J. Define on M_k, $R_{k,m} = (I - K_k^m) A_k^{-1}$ and recall that we have assumed in this discussion that $K_k = K_k^*$. Then, for $u \in M_k$,

$$(R_k u, u) = ((I - K_k) A_k^{-1} u, u) \le ((I - K_k^m) A_k^{-1} u, u) = (R_{k,m} u, u).$$

Hence A.4 is satisfied by $R_{k,m}$ if it is satisfied by R_k. Also A.5 is satisfied with $\theta = 1$. Thus we see that the previous results for $m = 1$, $p = 1$ are valid for $m \ge 1$, $p \ge 1$ when $K_k = K_k^*$ and $\sigma(K_k) \subset [0,1]$.

None of the results thus far obtained showed any improvement when the number of smoothing steps is increased. One very interesting result, which, in fact, provided the first uniform estimate for the V–cycle algorithm, was given by Braess and Hackbusch. This result also showed that, under certain circumstances, the contraction number improved as m is increased. This is contained in the following.

THEOREM 3.6. *Let B_k be defined by Algorithm II with $p = 1$. Assume that A.7 holds with $\alpha = 1$ and $K_k = I - \lambda_k^{-1} A_k$. Suppose that $\lambda_k / \lambda_{k-1} \le C$ for all k. Then there is a constant $C_1 > 0$ such that for $\delta = C_1 (2m + C_1)^{-1}$*

(3.28) $$0 \le A((I - B_k A_k) v, v) \le \delta A(v, v),$$

for all $v \in M_k$.

Proof: Recall the recurrence relation (3.27) on M_k

$$I - B_k A_k = K_k^m[(I - P_{k-1}) + (I - B_{k-1}A_{k-1})P_{k-1}]K_k^m.$$

Thus

$$A((I - B_k A_k)u, u) = A((I - P_{k-1})K_k^m u, K_k^m u)$$
$$+ A((I - B_{k-1}A_{k-1})P_{k-1}K_k^m u, P_{k-1}K_k^m u)$$
$$\geq 0$$

by induction.

Set $K_k^m u = \tilde{u}$. Then

$$A((I - B_k A_k)u, u) = A((I - P_{k-1})\tilde{u}, \tilde{u}) + A((I - B_{k-1}A_{k-1})P_{k-1}\tilde{u}, P_{k-1}\tilde{u}).$$

We prove the upper estimate by induction. For $k = 1$, $I - B_1 A_1 = 0$. Assume that it is true for $\delta = C_1(2m + C_1)^{-1}$ and $k - 1$. We shall see that C_1 may be chosen, independently of m and k such that (3.28) is also true for the same δ.

Using the induction hypothesis

$$A((I - B_k A_k)u, u) \leq A((I - P_{k-1})\tilde{u}, \tilde{u}) + \delta A(P_{k-1}\tilde{u}, P_{k-1}\tilde{u})$$
$$= (1 - \delta)A((I - P_{k-1})\tilde{u}, \tilde{u}) + \delta A(\tilde{u}, \tilde{u}).$$

We now use A.7 with $\alpha = 1$. Then

$$A((I - P_{k-1})\tilde{u}, \tilde{u}) \leq ((I - P_{k-1})\tilde{u}, A_k \tilde{u})$$
$$\leq \|(I - P_{k-1})\tilde{u}\| \, \|A_k \tilde{u}\|$$
$$\leq C\lambda_k^{-1/2}(A((I - P_{k-1})\tilde{u}, (I - P_{k-1})\tilde{u}))^{1/2}\|A_k \tilde{u}\|.$$

This implies that
$$A((I - P_{k-1})\tilde{u}, \tilde{u}) \leq C_1 \frac{\|A_k \tilde{u}\|^2}{\lambda_k}.$$

Hence

$$A((I - B_k A_k)u, u) \leq C_1(1 - \delta)\frac{\|A_k \tilde{u}\|^2}{\lambda_k} + \delta A(\tilde{u}, \tilde{u}).$$

Now

$$\frac{\|A_k \tilde{u}\|^2}{\lambda_k} = A\left(\frac{A_k}{\lambda_k} K_k^m u, K_k^m u\right)$$

$$= A((I - K_k)K_k^m u, K_k^m u)$$

$$= A((I - K_k)K_k^{2m} u, u).$$

Since $K_k = K_k^*$ and $\sigma(K_k) \subset [0, 1]$, it follows that

$$A((I - K_k)K_k^{2m} u, u) \leq \frac{1}{2m}A((I - K_k^{2m})u, u) = \frac{1}{2m}[A(u, u) - A(K_k^m u, K_k^m u)].$$

The last inequality follows from the spectral representation of K_k and the elementary inequality

$$(1 - z)z^{2m} \leq \frac{1}{2m}(1 - z)\sum_{i=0}^{2m-1} z^i = \frac{1}{2m}(1 - z^{2m})$$

for $0 \leq z \leq 1$. Thus

$$A((I - B_k A_k)u, u) \leq \frac{C_1}{2m}(1 - \delta)[A(u, u) - A(K_k^m u, K_k^m u)] + \delta A(K_k^m u, K_k^m u).$$

Now $\delta = C_1(2m + C_1)^{-1}$ so that

$$\delta = \frac{C_1}{2m}(1 - \delta).$$

Thus

$$A((I - B_k A_k)u, u) \leq \delta A(u, u)$$

which proves Theorem 3.6.

Bibliographical Notes

The first papers on multigrid were in the 1960's. Fedorenko first introduced the idea [84] in the context of finite differences. Then Bakhvalov [10] gave an analysis of the method of Fedorenko.

In the early 1970's Brandt studied this method and was a major advocate of it as a powerful tool for solving problems associated with elliptic equations. His work was extremely valuable and he had a lot to do with the popularization of the multigrid techniques (cf. [53]). A very important contribution to the theory was made by Bank and Dupont in [12] where they presented a rigorous treatment of multigrid as an optimal way to solve equations coming from finite element discretizations of elliptic problems. They assumed nested spaces and proved convergence of the W–cycle with sufficiently many smoothing steps, assuming some regularity in the continuous problem. This paper led to many subsequent works. They proved the part of Theorem 4.3 of the next section (in the nested case with the same form on all levels) with m sufficiently large.

Braess and Hackbusch [19] were the first to prove that the V–cycle converges uniformly with respect to the number of levels. This proof required "full elliptic regularity". It, however, led the way to the proofs that the W–cycle gave a uniform reduction under reduced regularity assumptions. The work of [32] and [15] contain proofs of such a result. In [32] and [70] the first estimates were given for the convergence of the V–cycle with less than full regularity, but these estimates showed a deterioration depending on the lack of full regularity. The paper [32] introduced the variable V–cycle and proved uniform estimates with some (arbitrary amount of) regularity.

The paper [45] introduced the additive multilevel preconditioner which proved to be a valuable tool in study of the general theory. The paper by King [105], in which he introduced a related two level preconditioner, had a strong influence on this work. In [45] a version of Theorem 3.1 was given, but the estimates there were not sharp and deteriorated with the number of levels. In this paper the technique for proving that very general mesh refinements could be included was first introduced. Oswald [135] gave uniform estimates for the case of piecewise linear functions on triangles using approximation theoretic techniques in Besov spaces. Independently Zhang [169] essentially proved, in this case, an estimate of the type A.2. This gave

56

rise to a uniform upper bound for the additive preconditioner.

The paper [42] was the first to give a treatment of the "product type" algorithms. This led to [43] in which the first multigrid results were given and specific regularity properties were not required. Theorem 3.3 is contained there. The mesh refinement techniques of [45] in the additive case were shown to hold in the standard multigrid algorithms.

The results of [169] and [135] motivated the papers [36] and [35] by Bramble and Pasciak in which uniform estimates were given for the multigrid V–cycle with only one smoothing step. These results as in Theorem 3.2 and Theorem 3.4 give an extension of the results of Braess and Hackbusch to the case of less than full regularity with one smoothing step. Theorem 3.6 of Braess and Hackbusch, although weaker than Theorems 3.2 or 3.4 in the case $m = 1$, shows improvement as m (the number of smoothings per step) is increased. This type of result led the way for the non-nested theory of Section 4.

4. Multigrid Algorithms with Nonnested Spaces and Varying Forms

In the previous section we always assumed that $M_1 \subset M_2 \subset \cdots \subset M_J \equiv M$. Thus the inner product $(\cdot, \cdot)$ and the form $A(\cdot, \cdot)$ defined on M are defined on M_k for any k. In some applications it is useful to not restrict ourselves to this case. Thus we want to allow the spaces M_k to be not necessarily nested and also we want to be able to define, for each k, a form $A_k(\cdot, \cdot)$ and an inner product $(\cdot, \cdot)_k$. We will follow as closely as possible the development in Section 3 but introducing now the more general framework. We recall that our goal is always to define an algorithm for the construction of an operator $B_J : M \rightarrow M$ which can be used to define a linear iterative process for solving (1.1) or a preconditioner for A.

The more general setup consists of several ingredients. We list them as follows.

1. Let $M_1, \ldots, M_J$ be J finite dimensional spaces with $M_J = M$.

2. Assume that we have $J - 1$ linear operators $I_k : M_{k-1} \rightarrow M_k$, $k = 2, \ldots, J$ which connect the spaces.

3. Assume that we have SPD forms $A_k(\cdot, \cdot)$ on $M_k \times M_k$, with $A_J(\cdot, \cdot) = A(\cdot, \cdot)$.

4. Assume that we have for each k an inner product $(\cdot, \cdot)_k$ on $M_k \times M_k$, with induced norm $\| \cdot \|_k$.

5. Define $A_k : M_k \rightarrow M_k$ by

$$(A_k w, \varphi)_k = A_k(w, \varphi), \text{ for all } \varphi \in M_k.$$

6. Define $P_{k-1} : M_k \rightarrow M_{k-1}$ and $Q_{k-1} : M_k \rightarrow M_{k-1}$ by

$$A_{k-1}(P_{k-1} w, \varphi) = A_k(w, I_k \varphi)$$

and

$$(Q_{k-1} w, \varphi)_{k-1} = (w, I_k \varphi)_k,$$

for all $\varphi \in M_{k-1}$.

58

7. Let $R_k : M_k \to M_k$ be a linear (smoothing) operator and in addition define

$$R_k^{(\ell)} = \begin{cases} R_k & \text{if } \ell \text{ is odd} \\ R_k^t & \text{if } \ell \text{ is even,} \end{cases}$$

where R_k^t is the adjoint with respect to the inner product $(\cdot, \cdot)_k$.

We are now ready to state our most general multigrid algorithm.

As before we will define an operator $B = B_J$ by means of an induction process.

<u>Algorithm III</u>: Let p be a positive integer and let $m(k)$ be a positive integer depending on k.

0) $B_1 = A_1^{-1}$.

For $g \in M_k$ we define $B_k g$ in terms of B_{k-1}, $2 \le k \le J$.

1) $x^0 = 0$, $q^0 = 0$.

2) Define x^ℓ, $\ell = 1, \ldots, m(k)$ by

$$x^\ell = x^{\ell-1} + R_k^{(\ell+m(k))}(g - A_k x^{\ell-1}).$$

3) $y^{m(k)} = x^{m(k)} + I_k q^p$, where q^i for $i = 1, \ldots, p$ is defined by

$$q^i = q^{i-1} + B_{k-1}[Q_{k-1}(g - A_k x^{m(k)}) - A_{k-1} q^{i-1}].$$

4) Define y^ℓ for $\ell = m(k)+1, \ldots, 2m(k)$ by

$$y^\ell = y^{\ell-1} + R_k^{(\ell+m(k))}(g - A_k y^{\ell-1}).$$

5) $B_k g = y^{2m(k)}$.

We want as before to derive a recurrence relation for the operator $I - B_k A_k$. In order to do this we need a little more notation. Recalling that R_k is not necessarily symmetric, set

$$\tilde{K}_k^{(m)} = \begin{cases} (K_k^* K_k)^{m/2} & \text{if } m \text{ is even} \\ (K_k^* K_k)^{\frac{m-1}{2}} K_k^* & \text{if } m \text{ is odd,} \end{cases}$$

with

$$K_k = I - R_k A_k \text{ and } K_k^* = I - R_k^t A_k.$$

From the definition of B_k the following two–level recurrence relation for $I - B_k A_k$ is easily derived as was done for Algorithms I and II.

(4.1) $I - B_k A_k = (\tilde{K}_k^{(m(k))})^* [I - I_k P_{k-1} + I_k (I - B_{k-1} A_{k-1})^p P_{k-1}] \tilde{K}_k^{(m(k))}.$

Note that $A_{k-1} P_{k-1} = Q_{k-1} A_k$ still holds. Note also that in the case discussed in Section 3 in which $M_k \subset M_{k+1}$, $A_k(\cdot, \cdot) = A(\cdot, \cdot)$ and $I_k = I$ the operators P_k and Q_k are projectors. Here P_k and Q_k are not necessarily projectors. In the case in which the spaces are nested and all of the forms are defined in terms of the form on M_J we have

(4.2) $$A_k(I_k u, I_k u) = A_{k-1}(u, u).$$

This is the so–called "variational assumption" and we call this case the nested–inherited case. The term inherited is used since, once we are given the operators I_k, all of the forms come from $A(\cdot, \cdot)$.

We will next introduce a weaker condition than (4.2).

A.9:

$$A_k(I_k u, I_k u) \leq A_{k-1}(u, u), \quad \text{for all } u \in M_{k-1}.$$

PROPOSITION 4.1. *A.9 holds if and only if*

$$A_{k-1}(P_{k-1} u, P_{k-1} u) \leq A_k(u, u), \quad \text{for all } u \in M_k$$

holds.

This follows easily from the Cauchy–Schwarz inequality.

PROPOSITION 4.2. *A.9 holds if and only if*

$$A_k((I - I_k P_{k-1}) u, u) \geq 0, \quad \text{for all } u \in M_k$$

holds.

This is obvious from Proposition 4.1 and the definitions of I_k and P_{k-1}.

60

PROPOSITION 4.3. *Let B_k be defined by Algorithm III. Then A.9 implies that*

$$A_k((I - B_k A_k)u, u) \geq 0, \text{ for all } u \in M_k.$$

Proof: Since $A_k(I_k P_{k-1} u, v) = A_{k-1}(P_{k-1} u, P_{k-1} v)$, $I_k P_{k-1}$ is symmetric with respect to $A_k(\cdot, \cdot)$. Hence by induction, since $I - B_1 A_1 = 0$, it follows from (4.1) that $I - B_k A_k$ is symmetric with respect to $A_k(\cdot, \cdot)$ and positive semidefinite.

One of our aims is to study the operator $I - B_k A_k$ and provide conditions under which, when A.9 holds, we can estimate δ_k between zero and one such that

$$0 \leq A_k((I - B_k A_k)u, u) \leq \delta_k A_k(u, u), \text{ for all } u \in M_k.$$

Or, if A.9 does not hold, we want to establish conditions under which we can estimate δ_k such that

$$|A_k((I - B_k A_k)u, u)| \leq \delta_k A_k(u, u)$$

holds.

In order to achieve our goals we will need to make some further assumptions.

The next condition is

A.10: For some α with $0 < \alpha \leq 1$ there exists C_α independent of k such that

$$|A_k((I - I_k P_{k-1})u, u)| \leq C_\alpha^2 \left(\frac{\|A_k u\|_k^2}{\lambda_k} \right)^\alpha (A_k(u, u))^{1-\alpha}.$$

In order to get a feeling for this condition we will examine it for a moment in the nested–inherited case. Note that in this case the condition

(4.3) $\qquad (A_k^{1-\alpha}(I - P_{k-1})u, (I - P_{k-1})u) \leq C \lambda_k^{-\alpha} A(u, u), \text{ for all } u \in M_k$

implies A.10. The inequality (4.3) is commonly referred to in the case of elliptic boundary value problems as "elliptic regularity pickup". That A.10 follows from (4.3) is easily seen as follows. For $u \in M_k$,

$$
\begin{aligned}
A((I - P_{k-1})u, u) &= ((I - P_{k-1})u, A_k u) \\
&\leq (A_k^{1-\alpha}(I - P_{k-1})u, (I - P_{k-1})u)^{1/2} (A_k^{1+\alpha} u, u)^{1/2} \\
&\leq C^{1/2} \lambda_k^{-\alpha/2} A((I - P_{k-1})u, u)^{1/2} (A_k^{1+\alpha} u, u)^{1/2}.
\end{aligned}
$$

Now

$$(A_k^{1+\alpha} u, u) \leq (A_k u, A_k u)^{\alpha} (A_k u, u)^{1-\alpha}.$$

Hence

$$A((I - P_{k-1})u, u) \leq C\lambda_k^{-\alpha} \|A_k u\|_k^{2\alpha} A(u, u)^{1-\alpha}.$$

In the more general case being considered in this section the assumption on the smoother is the same as before, since A.4 is a condition local to M_k and hence does not depend on whether or not the M_k's are nested.

The first result that we give concerns the V–cycle. It is a generalization of the Braess–Hackbusch result, Theorem 3.6, in that we allow nonnested spaces and non-inherited forms and, in addition we do not require that $\alpha = 1$ in A.10.

THEOREM 4.1. Let B_k be defined by Algorithm III. Assume that A.4, A.9 and A.10 hold. Let $p = 1$ and $m(k) = m$ for all k. Then

$$(4.4) \qquad 0 \leq A_k((I - B_k A_k)u, u) \leq \delta_k A_k(u, u), \quad \text{for all } u \in M_k$$

with

$$\delta_k = \frac{Mk^{\frac{1-\alpha}{\alpha}}}{Mk^{\frac{1-\alpha}{\alpha}} + m^{\alpha}}$$

for M large enough and independent of k.

Proof: Using (4.1), we find

$$A_k((I - B_k A_k)u, u) = A_k((I - I_k P_{k-1})\tilde{u}, \tilde{u}) + A_{k-1}((I - B_{k-1} A_{k-1})P_{k-1}\tilde{u}, P_{k-1}\tilde{u})$$

where $\tilde{u} = \tilde{K}_k^{(m)} u$. The lower estimate is just Proposition 4.3. We also obtain the upper bound inductively. For $k = 1$ there is nothing to prove. Assume that (4.4) is true for $k - 1$. Then

$$A_k((I - B_k A_k)u, u) \leq A_k((I - I_k P_{k-1})\tilde{u}, \tilde{u}) + \delta_{k-1} A_{k-1}(P_{k-1}\tilde{u}, P_{k-1}\tilde{u})$$

$$= A_k((I - I_k P_{k-1})\tilde{u}, \tilde{u}) + \delta_{k-1} A_k(I_k P_{k-1}\tilde{u}, \tilde{u})$$

$$= (1 - \delta_{k-1})A_k((I - I_k P_{k-1})\tilde{u}, \tilde{u}) + \delta_{k-1} A_k(\tilde{u}, \tilde{u}).$$

62

We first use A.10 and the arithmetic–geometric mean inequality. Then

$$A_k((I - I_k P_{k-1})\tilde{u}, \tilde{u}) \le C_\alpha^2 \left\{ \alpha\gamma_k \frac{\|A_k\tilde{u}\|_k^2}{\lambda_k} + (1 - \alpha)\gamma_k^{\frac{-\alpha}{1-\alpha}} A_k(\tilde{u}, \tilde{u}) \right\},$$

with $\gamma_k > 0$ to be chosen.

Now we proceed as in our proof of the Braess–Hackbusch result, Theorem 3.6. Suppose m is odd. By A.4

$$\frac{\|A_k\tilde{u}\|_k^2}{\lambda_k} \le C_R A_k(\overline{R}_k A_k \tilde{u}, \tilde{u})$$

$$\overline{R}_k A_k = I - K_k^* K_k.$$

Then $\tilde{u} = (K_k^* K_k)^{\frac{m-1}{2}} K_k^* u$. Set $\overline{K}_k = K_k K_k^*$. Then

$$\frac{\|A_k\tilde{u}\|_k^2}{\lambda_k} \le C_R A_k((I - \overline{K}_k)\overline{K}_k^m u, u)$$

$$\le \frac{C_R}{m} A_k((I - \overline{K}_k^m)u, u)$$

$$= \frac{C_R}{m}[A_k(u, u) - A_k(\tilde{u}, \tilde{u})].$$

A similar result holds for m even. We now put things together to get

$$A_k((I - B_k A_k)u, u) \le (1 - \delta_{k-1})C_\alpha^2 \frac{\alpha\gamma_k}{m} C_R[A_k(u, u) - A_k(\tilde{u}, \tilde{u})]$$

$$+ \left[(1 - \alpha)(1 - \delta_{k-1})C_\alpha^2 \gamma_k^{-\alpha/(1-\alpha)} + \delta_{k-1} \right] A_k(\tilde{u}, \tilde{u}).$$

The rest is to show that if we choose γ_k so that

$$(1 - \delta_{k-1})C_\alpha^2 \frac{\alpha\gamma_k}{m} C_R = (1 - \alpha)(1 - \delta_{k-1})C_\alpha^2 \gamma_k^{-\alpha/(1-\alpha)} + \delta_{k-1},$$

then they are less than or equal to δ_k with

$$\delta_k = \frac{Mk^{\frac{1-\alpha}{\alpha}}}{Mk^{\frac{1-\alpha}{\alpha}} + m^\alpha}.$$

The details are long and tedious and may be found in [32].

<u>Remark</u>: Notice that δ_k tends to one as k increases. This is in contrast to the results in the nested–inherited case under similar hypotheses, e.g. Theorem 3.4. The deterioration, however is only like a power of k and hence may not be too serious.

We now can prove some results for the W–cycle. The first involves a weakening of A.9 and yet yields a stronger result for the W–cycle than we have in Theorem 4.1 for the V–cycle. The weaker condition is the following.

A.11:

$$A_k(I_k u, I_k u) \leq 2A_{k-1}(u, u), \text{ for all } u \in M_{k-1}.$$

With this condition we can prove the following result for the W–cycle.

THEOREM 4.2. Let B_k be defined by Algorithm III. Suppose that A.4, A.11 and A.10 hold. Let $p = 2$ and $m(k) = m$ for all k. Then

$$|A_k((I - B_k A_k)u, u)| \leq \delta A_k(u, u), \text{ for all } u \in M_k,$$

with

$$\delta \leq \frac{M}{M + m^\alpha}.$$

Proof: We proceed by induction. For $k = 1$ there is nothing to prove. Assume that

$$|A_{k-1}((I - B_{k-1}A_{k-1})v, v)| \leq \delta A_{k-1}(v, v),$$

for all $v \in M_{k-1}$. Since $I - B_{k-1}A_{k-1}$ is symmetric with respect to $A_{k-1}(\cdot, \cdot)$, it follows that

$$A_{k-1}((I - B_{k-1}A_{k-1})^2 v, v) \leq \delta^2 A_{k-1}(v, v),$$

for all $v \in M_{k-1}$. Now, using (4.1), we have

$$\begin{aligned}
A_k((I - B_k A_k)u, u) &= A_k((I - I_k P_{k-1})\tilde{u}, \tilde{u}) \\
&\quad + A_{k-1}((I - B_{k-1}A_{k-1})^2 P_{k-1}\tilde{u}, P_{k-1}\tilde{u}) \\
&\leq A_k((I - I_k P_{k-1})\tilde{u}, \tilde{u}) + \delta^2 A_{k-1}(P_{k-1}\tilde{u}, P_{k-1}\tilde{u}) \\
&\leq (1 - \delta^2)|A_k((I - I_k P_{k-1})\tilde{u}, \tilde{u})| + \delta^2 A_k(\tilde{u}, \tilde{u}).
\end{aligned}$$

64

Now since A.11 is satisfied, we have that

$$-A_k((I - I_k P_{k-1})\tilde{u}, \tilde{u}) \le A_k(\tilde{u}, \tilde{u}).$$

From this we have, for any $\delta < 1$,

$$-A_k((I - B_k A_k)u, u) \le -A_k((I - I_k P_{k-1})\tilde{u}, \tilde{u})$$
$$\le (1 - \delta^2)[-A_k((I - I_k P_{k-1})\tilde{u}, \tilde{u})] + \delta^2 A_k(\tilde{u}, \tilde{u})$$
$$\le (1 - \delta^2)|A_k((I - I_k P_{k-1})\tilde{u}, \tilde{u})| + \delta^2 A_k(\tilde{u}, \tilde{u}).$$

This is the same bound as for $A_k((I - B_k A_k)u, u)$ and therefore

$$|A_k((I - B_k A_k)u, u)| \le (1 - \delta^2)|A_k((I - I_k P_{k-1})\tilde{u}, \tilde{u})| + \delta^2 A_k(\tilde{u}, \tilde{u}).$$

As in the proof of Theorem 4.1, using A.10, we see that for any $\gamma > 0$

$$|A_k((I - B_k A_k)u, u)| \le (1 - \delta^2)\frac{C_\alpha^2 \alpha \gamma}{m} C_R[A_k(u, u) - A_k(\tilde{u}, \tilde{u})]$$
$$+ \left[(1 - \alpha)(1 - \delta^2)C_\alpha^2 \gamma^{-\frac{\alpha}{1-\alpha}} + \delta^2\right] A_k(\tilde{u}, \tilde{u}).$$

Now choose γ to equate the coefficients and show that, for $\delta = \frac{M}{M+m^\alpha}$ with M large enough, they are less than or equal to δ. This shows that

$$|A_k((I - B_k A_k)u, u)| \le \delta A_k(u, u),$$

for $\delta = \frac{M}{M+m^\alpha}$ which proves Theorem 4.2.

We can next prove a version of the result of Bank and Dupont [12]. This is the following.

THEOREM 4.3. *Let B_k be defined by Algorithm III. Suppose that A.4 and A.10 hold. Let $p = 2$ and $m(k) = m$ for all k. Then for m sufficiently large there is a constant M which is independent of m such that*

$$|A_k((I - B_k A_k)u, u)| \le \delta A_k(u, u), \text{ for all } u \in M_k,$$

with

$$\delta \leq \frac{M}{M + m^\alpha}.$$

Proof: We proceed by induction. Again, as in the proof of Theorem 4.2 we assume that

$$A_{k-1}((I - B_{k-1}A_{k-1})^2 v, v) \leq \delta^2 A_{k-1}(v, v),$$

for all $v \in M_{k-1}$ and in exactly the same way as before show that

$$A_k((I - B_k A_k)u, u) \leq \delta A_k(u, u)$$

for $\delta = \frac{M}{M+m^\alpha}$. From the recurrence relation with $p = 2$ we have

$$-A_k((I - B_k A_k)u, u) \leq -A_k((I - I_k P_{k-1})\tilde{u}, \tilde{u}) \leq \frac{C}{m^\alpha} A_k(u, u).$$

Noticing that we may assume without loss that $M > C$, we now choose m such that $\frac{C}{m^\alpha} \leq \frac{M}{M+m^\alpha}$. This shows that

$$|A_k((I - B_k A_k)u, u)| \leq \delta A_k(u, u), \text{ for all } u \in M_k,$$

with

$$\delta \leq \frac{M}{M + m^\alpha}$$

which proves Theorem 4.3.

So far we have considered only the cases $p = 1$ and $p = 2$ with m, the number of smoothings the same on all levels. The case $p = 2$, of course, requires more work than does the case $p = 1$, but as we will see in typical situations in Section 10, the amount of work is not excessive. Another possible strategy, which has some advantages and which also does not increase the overall amount of work excessively, is a strategy in which the number of smoothings are increased as we proceed to coarser levels. We will call such a choice the variable V–cycle.

The next result is for this case. We make this precise as follows.

66

A.12: The number of smoothings, $m(k)$, increases as k decreases in such a way that

$$\beta_0 m(k) \leq m(k-1) \leq \beta_1 m(k)$$

with $1 < \beta_0 \leq \beta_1$.

Remark: One possible strategy would be to take $\beta_0 = \beta_1 = 2$. Thus the number of smoothings is doubled as we proceed to lower levels.

We now give the following result concerning the variable V–cycle.

THEOREM 4.4. *Let B_k be defined by Algorithm III. Suppose that A.4, A.9, A.10 and A.12 hold and that $p = 1$. Then*

$$0 \leq A_k((I - B_k A_k)u, u) \leq \delta_k A_k(u, u), \quad \text{for all } u \in M_k$$

with

$$\delta_k = \frac{M}{M + m(k)^\alpha}.$$

Remark: If e.g., $m(J) = 1$, we have

$$0 \leq A_J((I - B_J A_J)u, u) \leq \frac{M}{M+1} A_J(u, u), \quad \text{for all } u \in M_J.$$

The proof of Theorem 4.4 follows along the same lines as the last theorems. We omit it.

Now the condition A.9 is not always satisfied. What can we say without it? In such a case $A_k((I - B_k A_k)u, u)$ could become negative. Theorem 4.4 says that, with A.9,

$$(4.5) \qquad (1 - \delta_k) A_k(u, u) \leq A_k(B_k A_k u, u) \leq A_k(u, u).$$

However we used A.9 only to get the upper estimate. Notice that the lower inequality means that the operator $B_k A_k$ is positive definite. That this is true for any V–cycle algorithm is the next result.

67

THEOREM 4.5. *Let B_k be defined by Algorithm III. Assume A.4 only. Then, if $p = 1$, B_k is symmetric and positive definite.*

Proof: This is the same as $A_k(B_k A_k u, u) > 0$ when $u \neq 0$. We already saw that

$$(B_k A_k u, A_k u) = A_k((I - \overline{K}_k^{m(k)})u, u)$$
$$+ (B_{k-1} A_{k-1} P_{k-1} \tilde{K}_k^{(m(k))} u, A_{k-1} P_{k-1} \tilde{K}_k^{(m(k))} u)_{k-1}$$

where

$$\overline{K}_k = \begin{cases} K_k^* K_k & m \text{ even} \\ K_k K_k^* & m \text{ odd.} \end{cases}$$

The symmetry and positive definiteness now follow by induction using A.4 and the fact that $\sigma(\overline{K}_k) \subset [0, 1)$.

Now with this observation we do not need the upper estimate in (4.5) in order that B_k be useful. What we really want is to see that B_k is a good preconditioner, i.e., we want to estimate numbers η_0 and η_1 such that

(4.6) $\qquad \eta_0 A_k(u, u) \leq A_k(B_k A_k u, u) \leq \eta_1 A_k(u, u)$, for all $u \in M_k$.

We have the following:

THEOREM 4.6. *Let B_k be defined by Algorithm III. Assume that A.4, A.10 and A.12 hold and that $p = 1$. Then (4.6) holds with $\eta_0 \geq \frac{m(k)^a}{M + m(k)^a}$ and $\eta_1 \leq \frac{M + m(k)^a}{m(k)^a}$ for some $M > 0$.*

Proof: By (4.5), $1 - \delta_k \leq \eta_0$, where

$$\delta_k = \frac{M}{M + m(k)^a}.$$

Hence the bound for η_0 follows. In order to prove the estimate for η_1 we first prove the following.

LEMMA 4.1. *Let $p = 1$ and suppose that $\bar{\delta}_i$, $i = 2, \ldots, k$, satisfies*

$$-A_i((I - I_i P_{i-1})\tilde{u}_i, \tilde{u}_i) \leq \bar{\delta}_i A_i(u, u), \text{ for all } u \in M_i,$$

68

with $\tilde{u}_i = \tilde{K}_i^{(m(i))} u$. Then

$$\eta_1 \leq \prod_{i=2}^{k} (1 + \overline{\delta}_i).$$

Proof: Set $\tau_1 = 1$ and $\tau_k = \prod_{i=2}^{k}(1 + \overline{\delta}_i)$, $k \geq 2$. Then we need to show that

$$A_k(B_k A_k u, u) \leq \tau_k A_k(u, u)$$

or

(4.7) $$-A_k((I - B_k A_k)u, u) \leq (\tau_k - 1)A_k(u, u).$$

Again we proceed by induction. For $k = 1$ there is nothing to prove. Assume (4.7) is true for $k - 1$. Then

$$-A_k((I - B_k A_k)u, u)$$
$$= -A_k((I - I_k P_{k-1})\tilde{u}_k, \tilde{u}_k) - A_{k-1}((I - B_{k-1} A_{k-1})P_{k-1}\tilde{u}_k, P_{k-1}\tilde{u}_k)$$
$$\leq -A_k((I - I_k P_{k-1})\tilde{u}_k, \tilde{u}_k) + (\tau_{k-1} - 1)A_{k-1}(P_{k-1}\tilde{u}_k, P_{k-1}\tilde{u}_k)$$
$$\leq -A_k((I - I_k P_{k-1})\tilde{u}_k, \tilde{u}_k) - (\tau_{k-1} - 1)A_k((I - I_k P_{k-1})\tilde{u}_k, \tilde{u}_k)$$
$$\qquad + (\tau_{k-1} - 1)A_k(u, u)$$
$$\leq [\overline{\delta}_k + (\tau_{k-1} - 1)(1 + \overline{\delta}_k)]A_k(u, u)$$
$$= [\tau_{k-1}(1 + \overline{\delta}_k) - 1]A_k(u, u) = (\tau_k - 1)A_k(u, u).$$

This proves the lemma.

Now to estimate η_1 we estimate $\overline{\delta}_k$ using A.10:

$$-A_k((I - I_k P_{k-1})\tilde{u}_k, \tilde{u}_k) \leq C_\alpha^2 (\lambda_k^{-1} \|A_k \tilde{u}_k\|_k^2)^\alpha (A_k(\tilde{u}_k, \tilde{u}_k))^{1-\alpha}.$$

As before

$$\frac{\|A_k \tilde{u}_k\|_k^2}{\lambda_k} \leq \frac{C}{m(k)} A_k((I - \overline{K}_k^{m(k)})u, u) \leq \frac{C}{m(k)} A_k(u, u).$$

Hence, since $A_k(\tilde{u}_k, \tilde{u}_k) \leq A_k(u, u)$, we have that

$$-A_k((I - I_k P_{k-1})\tilde{u}_k, \tilde{u}_k) \leq \frac{C}{m(k)^\alpha} A_k(u, u).$$

69

Thus

$$\eta_1 \leq \prod_{k=2}^{J} \left(1 + \frac{C}{m(k)^\alpha}\right).$$

It follows from A.12 that, for some M,

$$\prod_{k=2}^{J} \left(1 + \frac{C}{m(k)^\alpha}\right) \leq 1 + \frac{M}{m(J)^\alpha} = \frac{M + m(J)^\alpha}{m(J)^\alpha}.$$

Bibliographical Notes

This section is a condensation of the results of Bramble, Pasciak and Xu [46].

Condition A.9, at first, may seem somewhat artificial, since, in the nested–inherited case A.9 holds with equality. However, Sections 5, 6 and 8 of [46] contain natural examples in which A.9 is satisfied without equality.

The result of Bank and Dupont [12], Theorem 4.3, is not a strong result by current standards when applied in the nested–inherited case as they did. However it was an important development in the analysis of multigrid algorithms. Nevertheless, there are situations in which this type of result seems to be the only one within reach c.f. [156].

Perhaps the most novel of the results of this section is that of Theorem 4.6 which shows the robustness of the variable V–cycle in that, with any amount of regularity, it always gives rise to a uniform preconditioner. Applications will be mentioned in a later section.

5. Construction of Classes of Smoothers

In Section 3 a generic smoother R_k was introduced and was assumed to satisfy certain conditions. This section is devoted to the construction of two classes of smoothers which satisfy the relevant conditions. We will see in Section 10 that these classes of smoothers are constructed in such a way that the multigrid algorithms are easily implemented. These classes include the block Jacobi and Gauss–Seidel smoothers.

Recall that $R_k : M_k \to M_k$ and $K_k = I - R_k A_k$, $K_k^* = I - R_k^t A_k$. One condition which we would like R_k to satisfy is A.4. We state it again here for conveience.

A.4: There exists $C_R \geq 1$ and independent of k such that

$$\frac{\|u\|^2}{\lambda_k} \leq C_R(\overline{R}_k u, u)$$

where

$$\overline{R}_k = (I - K_k^* K_k) A_k^{-1}.$$

The equivalent form of A.4 is

$$(A_k K_k u, K_k u) \leq (A_k K_{k,\omega} u, u)$$

where $K_{k,\omega} = I - \omega \lambda_k^{-1} A_k$ with $\omega = 1/C_R$. Recall also that we required often that a scaling condition be satisfied. We also state it again here for conveience.

A.5: there exists θ, between zero and two, such that

$$(A_k T_k v, T_k v) \leq \theta(A_k T_k v, v),$$

where $T_k = R_k A_k$.

We shall construct, in simple ways, smoothing operators R_k satisfying A.4 and A.5. This will be done by decomposing the space M_k into subspaces M_k^i, $i = 1, \ldots, \ell$, with

$$M_k = \sum_{i=1}^{\ell} M_k^i.$$

This decomposition may or may not be a direct sum, i.e.,

$$u = \sum_{i=1}^{\ell} u_i$$

where $u_i \in M_k^i$ may not be unique. Define $A_{k,i} : M_k^i \to M_k^i$ and $P_k^i, Q_k^i : M_k \to M_k^i$ by

$$(A_{k,i}v, \chi) = (A_k v, \chi),$$

$$(A_k P_k^i v, \chi) = (A_k v, \chi) \text{ and } (Q_k^i v, \chi) = (v, \chi),$$

for all $\chi \in M_k^i$. The first smoother is defined as an "additive smoother":

(5.1)
$$R_k = \gamma \sum_{i=1}^{\ell} A_{k,i}^{-1} Q_k^i.$$

In (5.1) γ is a positive scaling factor to be chosen. Define the "interaction matrix"

$$\kappa_{im} = \begin{cases} 0 & \text{if } P_k^i P_k^m = 0 \\ 1 & \text{otherwise} \end{cases}$$

and set

$$n_k = \max_i \sum_{m=1}^{\ell} \kappa_{im}$$

(maximum row sum).

We assume two simple hypotheses on the subspaces, namely

(1) There is a number C_1 such that $n_k \leq C_1$ with C_1 independent of k.

(2) There exists C_0 independent of k such that for each $u \in M_k$ there is a decomposition

$$u = \sum_{i=1}^{\ell} u_i, \quad u_i \in M_k^i$$

satisfying

$$\sum_{i=1}^{\ell} \|u_i\|^2 \leq C_0 \|u\|^2.$$

Under these two hypotheses we can prove the following.

72

THEOREM 5.1. *Let R_k be defined by (5.1) and suppose that (1) and (2) above are satisfied. Let $\theta \in (0,2)$ and set $\gamma = \theta/C_1$. Then A.4 and A.5 are satisfied with $C_R = C_0 C_1/\theta(2-\theta)$.*

We first prove the following simple lemma.

LEMMA 5.1. *Let n_k be as above and u_i and $v_i \in M_k^i$, $i = 1, \dots, \ell$. Then*

$$\left(\sum_{i,m=1}^{\ell} |(A_k u_i, v_m)| \right)^2 \leq n_k^2 \sum_{i=1}^{\ell} (A_k u_i, u_i) \sum_{m=1}^{\ell} (A_k v_m, v_m).$$

Proof: Note that

$$\left(\sum_{i,m=1}^{\ell} |(A_k u_i, v_m)| \right)^2 = \left(\sum_{i,m=1}^{\ell} \kappa_{im} |(A_k u_i, v_m)| \right)^2$$

$$\leq \sum_{i,m=1}^{\ell} \kappa_{im} (A_k u_i, u_i) \sum_{i,m=1}^{\ell} \kappa_{im} (A_k v_m, v_m)$$

$$\leq n_k^2 \sum_{i=1}^{\ell} (A_k u_i, u_i) \sum_{m=1}^{\ell} (A_k v_m, v_m).$$

Proof of Theorem 5.1: We first show, noting that $R_k = R_k^t$, that

(5.2) $$\frac{\|u\|^2}{\lambda_k} \leq C_0 \gamma^{-1} (R_k u, u).$$

Let $u = \sum_{i=1}^{\ell} u_i$ satisfy (2). Then

$$\|u\|^2 = \sum_{i=1}^{\ell} (u_i, Q_k^i u) \leq \left(\sum_{i=1}^{\ell} \|u_i\|^2 \right)^{1/2} \left(\sum_{i=1}^{\ell} \|Q_k^i u\|^2 \right)^{1/2}$$

$$\leq C_0^{1/2} \|u\| \left(\sum_{i=1}^{\ell} \|Q_k^i u\|^2 \right)^{1/2}.$$

Hence

(5.3) $$\|u\|^2 \leq C_0 \sum_{i=1}^{\ell} \|Q_k^i u\|^2 \leq C_0 \lambda_k \sum_{i=1}^{\ell} (A_{k,i}^{-1} Q_k^i u, u) = C_0 \lambda_k \gamma^{-1} (R_k u, u),$$

which is (5.2). We next want to show that A.5 holds if $\gamma = \theta/C_1$. First note that T_k is SPD (with respect to $(A_k \cdot, \cdot)$). Hence it is sufficient to show that

$$(5.4) \qquad (A_k T_k v, v) \leq \theta(A_k v, v) \text{ if } \gamma = \theta/C_1.$$

That (5.4) implies A.5 is easily seen by taking $v = T_k^{1/2} u$ and noticing that $T_k^{1/2}$ is also symmetric with respect to $(A_k \cdot, \cdot)$ and commutes with T_k. We now show (5.4). By definition

$$T_k = R_k A_k = \gamma \sum_i A_{k,i}^{-1} Q_k^i A_k = \gamma \sum_i A_{k,i}^{-1} A_{k,i} P_k^i = \gamma \sum_i P_k^i.$$

Hence, taking $v_i = P_k^i v$, we have

$$(A_k T_k v, v) = \gamma \sum_{i=1}^{\ell} (A_k v_i, v) = \gamma \sum_{i=1}^{\ell} (A_k v_i, v_i).$$

We note that

$$\sum_{i=1}^{\ell} (A_k v_i, v_i) = \sum_{i=1}^{\ell} (A_k v_i, v) \leq \left(A_k \sum_{i=1}^{\ell} v_i, \sum_{m=1}^{\ell} v_m \right)^{1/2} (A_k v, v)^{1/2}$$

$$\leq \left(\sum_{i,m=1}^{\ell} |(A_k v_i, v_m)| \right)^{1/2} (A_k v, v)^{1/2}.$$

By Lemma 5.1, it follows that

$$\sum_{i=1}^{\ell} (A_k v_i, v_i) \leq C_1 (A_k v, v).$$

Thus

$$(A_k T_k v, v) \leq \gamma C_1 (A_k v, v) = \theta(A_k v, v).$$

We still need to verify A.4. We will show that

$$(5.5) \qquad (\overline{R}_k u, u) \geq (2 - \theta)(R_k u, u).$$

74

Combining this with (5.2), we get A.4 with $C_R = C_0 \gamma^{-1}(2-\theta)^{-1} = C_0 C_1 [\theta(2-\theta)]^{-1}$. Now (5.4) can be written as $(A_k R_k A_k v, v) \leq \theta(A_k v, v)$. It follows from Lemma 3.2 that this is equivalent to

$$(5.6) \qquad (R_k A_k R_k v, v) \leq \theta(R_k v, v).$$

Now $\overline{R}_k = [I - (I - R_k A_k)(I - R_k A_k)]A_k^{-1} = 2R_k - R_k A_k R_k$, which, together with (5.6) gives (5.5).

The operator R_k defined by (5.1) can be thought of as a Jacobi–type smoother.

Another important class of smoothers is the class of multiplicative smoothers based on the same subspace decomposition (Gauss–Seidel).

Multiplicative Algorithm: For $f \in M_k$, define $R_k f \in M_k$ as follows:

1. Set $v_0 = 0$

2. For $i = 1, \dots, \ell$ define

$$v_i = v_{i-1} + A_{k,i}^{-1} Q_k^i (f - A_k v_{i-1})$$

3. $R_k f = v_\ell$.

Now $A_{k,i} P_k^i = Q_k^i A_k$. Hence

$$K_k = I - R_k A_k = (I - P_k^\ell) \cdots (I - P_k^1).$$

To see this take $f = A_k x$. Then

$$x - v_i = x - v_{i-1} - (A_{k,i}^{-1} Q_k^i A_k)(x - v_{i-1}) = (I - P_k^i)(x - v_{i-1}).$$

Hence

$$(I - R_k A_k)x = x - v_\ell = (I - P_k^\ell)(x - v_{\ell-1}) = (I - P_k^\ell) \cdots (I - P_k^1)x.$$

Now we can prove the following.

75

THEOREM 5.2. *Let R_k be defined by the multiplicative algorithm and assume (1) and (2) with C_0 and C_1 as before. Then A.4 and A.5 are satisfied with $C_R = 2C_0(1 + C_1^2)$ and $\theta = \frac{2C_1}{C_1+1}$.*

For A.4 we prove the alternative statement, (3.14),

(5.7) $$(A_k K_k u, K_k u) \leq (A_k K_{k,\omega} u, u)$$

with

$$K_{k,\omega} = I - \omega \lambda_k^{-1} A_k, \quad \omega = 1/C_R.$$

Set $E_i^k = (I - P_k^i) \cdots (I - P_k^1)$ and $E_0^k = I$. Note that $K_k = E_\ell^k$ and that $K_k^* = (I - P_k^1) \cdots (I - P_k^\ell)$.

We use the same techniques as in the multigrid proofs. Clearly

$$E_{i-1}^k - E_i^k = P_k^i E_{i-1}^k$$

and hence

$$I - E_i^k = \sum_{m=1}^{i} P_k^m E_{m-1}^k.$$

Thus

$$(A_k E_{i-1}^k v, E_{i-1}^k v) - (A_k E_i^k v, E_i^k v) = (A_k P_k^i E_{i-1}^k v, E_{i-1}^k v).$$

Summing gives

(5.8) $$(A_k v, v) - (A_k E_\ell^k v, E_\ell^k v) = \sum_{i=1}^{\ell} (A_k P_k^i E_{i-1}^k v, E_{i-1}^k v).$$

Rewrite (5.7) as

(5.9) $$\omega \lambda_k^{-1} \|A_k v\|^2 \leq (A_k v, v) - (A_k E_\ell^k v, E_\ell^k v).$$

We shall prove (5.9). By (5.3)

$$\|u\|^2 \leq C_0 \sum_{i=1}^{\ell} \|Q_k^i u\|^2$$

76

and hence

$$\|A_k v\|^2 \le C_0 \sum_{i=1}^{\ell} \|Q_k^i A_k v\|^2 = C_0 \sum_{i=1}^{\ell} (A_{k,i} P_k^i v, A_{k,i} P_k^i v) \le C_0 \lambda_k \sum_{i=1}^{\ell} (A_k P_k^i v, P_k^i v).$$

Thus (5.9) will follow from (5.8) if we show that

(5.10)
$$\sum_{i=1}^{\ell} (A_k P_k^i v, P_k^i v) \le 2(1 + C_1^2) \sum_{i=1}^{\ell} (A_k P_k^i E_{i-1}^k v, E_{i-1}^k v).$$

To show this we write

$$\left(\sum_{i=1}^{\ell} (A_k P_k^i v, P_k^i v) \right)^2 = \left[\sum_{i=1}^{\ell} (A_k P_k^i v, P_k^i E_{i-1}^k v) + \sum_{i=1}^{\ell} \sum_{m=1}^{i-1} (A_k P_k^i v, P_k^m E_{m-1}^k v) \right]^2$$

$$\le 2 \left\{ \sum_{i=1}^{\ell} (A_k P_k^i v, P_k^i v) \sum_{i=1}^{\ell} (A_k P_k^i E_{i-1}^k v, E_{i-1}^k v) \right.$$

$$\left. + \left[\sum_{i,m=1}^{\ell} |(A_k P_k^i v, P_k^m E_{m-1}^k v)| \right]^2 \right\}.$$

By Lemma 5.1 this implies that

$$\left(\sum_{i=1}^{\ell} (A_k P_k^i v, P_k^i v) \right)^2 \le 2(1 + n_k^2) \sum_{i=1}^{\ell} (A_k P_k^i v, P_k^i v) \sum_{i=1}^{\ell} (A_k P_k^i E_{i-1}^k v, E_{i-1}^k v).$$

Thus (5.10) follows which implies (5.9).

We obtain the estimate for θ as follows. Recall that $T_k = R_k A_k = I - K_k = I - E_\ell^k$. Hence

$$(A_k T_k u, T_k u) = (A_k(I - E_\ell^k)u, (I - E_\ell^k)u) = \sum_{i,m=1}^{\ell} (A_k P_k^i E_{i-1}^k u, P_k^m E_{m-1}^k u).$$

Using Lemma 5.1

$$(A_k T_k u, T_k u) \le n_k \sum_{i=1}^{\ell} (A_k P_k^i E_{i-1}^k u, E_{i-1}^k u)$$

and from (5.8) we obtain

$$(A_k T_k u, T_k u) \le C_1((A_k u, u) - (A_k E_\ell^k u, E_\ell^k u))$$

$$= C_1((A_k u, u) - (A_k(I - T_k)u, (I - T_k)u)).$$

Thus

$$(A_k T_k u, T_k u) \leq C_1(2(A_k T_k u, u) - (A_k T_k u, T_k u))$$

which may be written as

$$(5.11) \qquad (A_k T_k u, T_k u) \leq \frac{2C_1}{C_1 + 1}(A_k T_k u, u).$$

We note here that (5.11) is the same as

$$A([(\theta/2)I - T_k]u, [(\theta/2)I - T_k]u) \leq (\theta/2)^2 A(u, u),$$

with $\theta = \frac{2C_1}{C_1+1}$. From this it easily follows that (5.11) is valid with T_k replaced by T_k^*. Finally, we note that by renaming the subspaces M_k^i, the same results hold for K_k replaced by K_k^*. This completes the proof of the theorem.

Bibliographical Notes

Various smoothers have been introduced in the literature. The ones most often introduced correspond to the point–Jacobi or Richardson type iteration (cf. [12]). Other authors have considered the Gauss–Seidel or symmetric Gauss–Seidel iteration as a smoother (cf. [14] and [15]). The treatment in this section is taken from [34] and includes point and block Jacobi and Gauss–Seidel as well as smoothers corresponding to overlapping domain decomposition preconditioners [73].

6. Application to Second Order Elliptic Problems

In this section we will show how the multigrid method applies to finite element approximations to second order elliptic boundary value problems. We shall consider domains with curved boundaries as well as the use of mesh refinements. For simplicity of exposition, we restrict ourselves to polygonal domains in the case of mesh refinements.

<u>Domains with curved boundaries</u>

We shall first consider the Dirichlet problem for domains $\Omega \subset R^2$ with curved boundaries. For simplicity we let $\partial\Omega$ be piecewise smooth and such that each smooth part of the boundary, Γ_i, is convex in the sense that the line joining any two points on Γ_i lies in $\overline{\Omega}$. We make this convexity assumption for simplicity. It is only technically more involved to treat the more general case and hence we do not do so here.

We now consider

$$(6.1) \qquad A(v,w) = \sum_{i,j=1}^{2} \int_{\Omega} a_{ij} \frac{\partial v}{\partial x_i} \frac{\partial w}{\partial x_j} dx + \int_{\Omega} avw\,dx$$

on $H_0^1(\Omega) \times H_0^1(\Omega)$, where the 2×2 symmetric matrix with entries a_{ij} is uniformly positive definite and the function a is non–negative.

The Dirichlet problem is thus: Given f, find $u \in H_0^1(\Omega)$ such that

$$(6.2) \qquad A(u,w) = (f,w), \text{ for all } w \in H_0^1(\Omega).$$

Once we define the subspace $M \subset H_0^1(\Omega)$, the Galerkin approximation is: Find $U_h \in M$ such that

$$(6.3) \qquad A(U_h,\chi) = (f,\chi), \text{ for all } \chi \in M.$$

79

We assume that a_{ij} and a are smooth enough so that the following a priori estimate is valid:

(6.4)
$$\|u\|_{1+\alpha} \leq C\|f\|_{-1+\alpha}$$

for some $\alpha \in (0, 1]$. The norm $\|\cdot\|_{1+\alpha}$ is the norm on the space defined by interpolation between the spaces $H_0^1(\Omega)$ and $H^2(\Omega)$. The space with negative index is defined by duality. That is

$$\|f\|_{-1+\alpha} = \sup_{\theta \in H_0^1(\Omega)} \frac{(f, \theta)}{\|\theta\|_{1-\alpha}},$$

where the norm $\|\cdot\|_{1-\alpha}$ is the norm on the space defined by interpolation between the spaces $L^2(\Omega)$ and $H_0^1(\Omega)$.

We start by defining $M = M_J$ in such a way that we have subspaces $M_1 \subset M_2 \subset \cdots \subset M_J = M \subset H_0^1(\Omega)$. Let T_1 be a coarse triangulation in Ω. We take certain points on the boundary (including the end–points of each Γ_i) and in the interior and connect them so that the resulting triangles are in Ω and no triangle has more than two points on $\partial\Omega$.

Now define T_k from T_{k-1} by first connecting the midpoints of the sides of each triangle with two vertices in Ω. If a triangle in T_{k-1} has two vertices on the boundary, we assume that they are connected by a smooth arc on the boundary and we take a new boundary point to be the midpoint of this arc. We then form four new triangles by connecting the midpoints of the resulting "curved triangle".

Note that in this construction not all triangles in T_{k-1} are the union of triangles of T_k. As a consequence, if we define the piecewise linear subspaces of $H_0^1(\Omega)$ in a natural way, we get a set of nonnested spaces $\overline{M}_k$, $k = 1, \ldots, J$, where the functions in $\overline{M}_k$ are zero outside of the triangles of T_k.

By the above construction we take $M \equiv M_J = \overline{M}_J$. We want to define nested spaces for use in our multigrid algorithms.

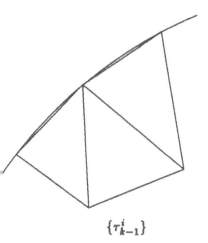
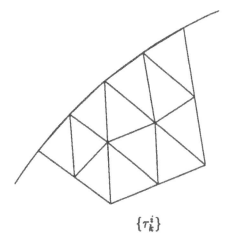

$$\{\tau^i_{k-1}\} \qquad\qquad\qquad \{\tau^i_k\}$$

Figure 6.1

The mesh refinement near the boundary.

Construction of M_k, $k < J$.

Let Ω^0_k, for $k < J$, be the interior of the union of the (closed) triangles of $\mathcal{T}_k$ which do not have vertices on $\partial\Omega$. The spaces M_k, for $k < J$, are defined by

$$(6.5) \qquad\qquad M_k = \{\varphi \in \overline{M}_k \mid supp(\varphi) \subset \overline{\Omega}^0_k\}.$$

Thus $M_k \subset M_{k+1}$. This is clear since each triangle in which some function in M_k is nonzero is the union of four triangles in which some functions in M_{k+1} are nonzero.

Having defined $M_1 \subset M_2 \subset \cdots \subset M_J$, we choose a smoother satisfying A.4 and A.5. In order to apply Theorem 3.4 we note that the condition on λ_k is satisfied since $\lambda_k \approx h_k^{-2}$. As in our first example h_k is the length of the longest side of all the triangles in $\mathcal{T}_k$ and, by construction, h_{k+1} is approximately equal to $\frac{1}{2}h_k$.

It remains to show that A.7 is satisfied; i.e., for some $\alpha \in (0, 1]$,

$$(6.6) \qquad\qquad (A^{1-\alpha}(I - P_k)v, (I - P_k)v) \le Ch_k^{2\alpha}A(v, v)$$

and

$$(6.7) \qquad\qquad (A_k^{1-\alpha}Q_kv, Q_kv) \le C(A^{1-\alpha}v, v),$$

81

for all $v \in M$. For this purpose we prove first the following:

LEMMA 6.1. *Let Ω^η denote the strip $\{x \in \Omega \mid dist(x, \partial\Omega) < \eta\}$ and $0 < s < 1/2$.
Then, for all $v \in H^{1+s}(\Omega)$,*

$$\|v\|_{1,\Omega^\eta} \leq C\eta^s \|v\|_{1+s}.$$

In addition for $v \in H_0^1(\Omega)$,

(6.8)
$$\|v\|_{\Omega^\eta} \leq C\eta \|v\|_1.$$

Proof: Let ω^η be a square of side length η. Then

(6.9)
$$\|w\|_{\omega^\eta}^2 \leq \eta^2/2 \int_{\omega^\eta} |\nabla w|^2 dx$$

holds for all w vanishing on one edge of ω^η. An inequality similar to (6.9) is also
true if w vanishes on, for example, an open segment of one of the sides.

Inequality (6.8) follows by covering the strip Ω^η with subregions Ω_i^η each of which
can be smoothly mapped onto w^η.

Now this, together with the trivial inequality

$$\|v\|_{\Omega^\eta} \leq \|v\|$$

yields, by Theorem B.4,

(6.10)
$$\|v\|_{\Omega^\eta} \leq C\eta^s \|v\|_s$$

for $v \in H_0^s(\Omega)$, with $s \in [0,1]$, $s \neq 1/2$, and an appropriate constant $C \geq 0$. But
by a result in [96] if $0 < s < 1/2$ then $H_0^s(\Omega) = H^s(\Omega)$ (provided $\partial\Omega$ is Lipschitz).
Thus applying (6.10) to the first derivative of v completes the proof of the lemma.

Now the inequalities (6.6) and (6.7) will follow from the Aubin–Nitsche duality
argument and interpolation, provided we prove that for $w \in H^{1+\alpha}(\Omega) \cap H_0^1(\Omega)$

(6.11)
$$\|(I - P_k)w\|_1 \leq C h_k^\alpha \|w\|_{1+\alpha},$$

and for $v \in H_0^1(\Omega)$

$$(6.12) \qquad\qquad A(Q_k v, Q_k v) \leq C A(v, v).$$

We postpone the proof that (6.11) and (6.12) imply (6.6) and (6.7) until the end of this section.

In order to prove (6.11) let $\chi \in M_k$ be the interpolant of w at the nodes of Ω_k^0 and $\overline{\chi} \in \overline{M}_k$ which interpolates w at the nodes of Ω. Then

$$A((I - P_k)w, (I - P_k)w) \leq A(w - \chi, w - \chi)$$
$$\leq C\{\|w - \overline{\chi}\|_1^2 + \|\overline{\chi} - \chi\|_{1,\Omega_k^0}^2 + \|w\|_{1,\Omega\setminus\Omega_k^0}^2\}.$$

Note that $\overline{\chi} - \chi \in \overline{M}_k$ and vanishes at all vertices except those on $\partial\Omega_k^0$. Thus

$$\|\overline{\chi} - \chi\|_{1,\Omega_k^0}^2 \leq C \sum_{x_i \in \partial\Omega_k^0} \overline{\chi}(x_i)^2 \leq C\|\overline{\chi}\|_{1,\Omega_k\setminus\Omega_k^0}^2$$
$$\leq C(\|w - \overline{\chi}\|_{1,\Omega_k\setminus\Omega_k^0}^2 + \|w\|_{1,\Omega_k\setminus\Omega_k^0}^2)$$
$$\leq C(h_k^{2\alpha}\|w\|_{1+\alpha}^2 + \|w\|_{1,\Omega\setminus\Omega_k^0}^2),$$

where we have used Lemma 6.1 with $\eta = h_k$ and the well known inequality

$$\|w - \overline{\chi}\|_1^2 \leq C h_k^{2\alpha}\|w\|_{1+\alpha}^2.$$

This proves (6.11).

We now prove (6.12). Let $\overline{Q}_k$ denote the L_2–projection onto $\overline{M}_k$. First, by using the interpolant in $\overline{M}_k$ we have that

$$\|(I - \overline{Q}_k)v\| \leq C h_k^{1+\alpha}\|v\|_{1+\alpha}$$

and we conclude by Theorem B.4 of operators that

$$(6.13) \qquad\qquad \|(I - \overline{Q}_k)v\|^2 \leq C\lambda_k^{-1}\|v\|_1^2, \ v \in H_0^1(\Omega).$$

Now let $\theta_k \in M_k$ be equal to $\overline{Q}_k v$ at the interior nodes of Ω_k^0. Then

$$\|(I - Q_k)v\|^2 \leq \|v - \theta_k\|^2 \leq 2(\|v - \theta_k\|_{\Omega_k^0}^2 + \|v - \theta_k\|_{\Omega\setminus\Omega_k^0}^2)$$
$$\leq C(\|(I - \overline{Q}_k)v\|_{\Omega_k^0}^2 + \|\overline{Q}_k v - \theta_k\|_{\Omega_k^0}^2 + \|v\|_{\Omega\setminus\Omega_k^0}^2).$$

83

Hence, by Lemma 6.1 and (6.13)

$$\|(I - Q_k)v\|^2 \le C\lambda_k^{-1}\|v\|_1^2 + C\|\overline{Q}_k v - \theta_k\|_{\Omega_k^0}^2.$$

Now $\overline{Q}_k v - \theta_k \in \overline{M}_k$ and vanishes at all vertices except those on $\partial\Omega_k^0$. Hence

$$\begin{aligned}
\|\overline{Q}_k v - \theta_k\|_{\Omega_k^0}^2 &\le Ch_k^2 \sum_{x_i \in \partial\Omega_k^0} (\overline{Q}_k v(x_i))^2 \\
&\le C\|\overline{Q}_k v\|_{\Omega_k \backslash \Omega_k^0}^2 \\
&\le C(\|\overline{Q}_k v - v\|_{\Omega_k \backslash \Omega_k^0}^2 + \|v\|_{\Omega_k \backslash \Omega_k^0}^2) \\
&\le C\lambda_k^{-1}\|v\|_1^2.
\end{aligned}$$

This shows that

(6.14) $$\|(I - Q_k)v\|^2 \le C\lambda_k^{-1}\|v\|_1^2.$$

We finally show that

$$\|Q_k v\|_1^2 \le C\|v\|_1^2 \text{ for } v \in H_0^1.$$

Following [50], we use the local L^2 projection; $Q_\tau : L_2(\tau) \to P_1(\tau)$ where τ is a triangle in Ω_k^0 and $P_1(\tau)$ is the space of piecewise linear functions on τ. Let $\hat{\tau}$ be a unit size reference triangle and $F : \hat{\tau} \to \tau$ the affine map taking $\hat{\tau}$ to τ. Set

$$\hat{u}(\hat{x}) = u(F(\hat{x})) \text{ for } \hat{x} \in \hat{\tau}.$$

Define $Q_{\hat{\tau}}$ as above. Then $\widehat{Q_\tau u} = Q_{\hat{\tau}}\hat{u}$. Now we can see that

(6.15) $$|\nabla Q_\tau u|_\tau \le C|\nabla u|_\tau.$$

Mapping to $\hat{\tau}$ shows that (6.15) is equivalent to

(6.16) $$|\nabla Q_{\hat{\tau}}\hat{u}|_{\hat{\tau}} \le C|\nabla \hat{u}|_{\hat{\tau}}.$$

To prove (6.16) we note that since $P_1(\hat{\tau})$ is finite dimensional $(\dim P_1(\hat{\tau}) = 3)$

$$\|\nabla Q_{\hat{\tau}}\hat{u}\|_{\hat{\tau}} \le C\|Q_{\hat{\tau}}\hat{u}\|_{\hat{\tau}} \le C\|\hat{u}\|_{\hat{\tau}}.$$

84

But $Q_\tau \gamma = \gamma$ for any constant γ. Hence, by the Poincaré inequality,

$$|\nabla Q_\tau \hat{u}|_{\hat{\tau}} \leq C \inf_\gamma \|\hat{u} + \gamma\|_{\hat{\tau}} \leq C |\nabla \hat{u}|_{\hat{\tau}}.$$

Mapping back to τ yields the result. Now let Θ_k be the piecewise linear function on Ω_k which is the local L_2-projection of v. Then

$$\|v - \Theta_k\|^2 \leq \|v - Q_k v\|^2 \leq C \lambda_k^{-1} \|v\|_1^2$$

and

$$\sum_{\tau \in \mathcal{T}_k} \|\Theta_k\|_{1,\tau}^2 \leq C \|v\|_1^2.$$

Thus

$$
\begin{aligned}
(6.17) \qquad A(Q_k v, Q_k v) &\leq C \sum_\tau \|Q_k v - \Theta_k\|_{1,\tau}^2 + C \|v\|_1^2 \\
&\leq C \lambda_k \|Q_k v - \Theta_k\|_{\Omega_k}^2 + C \|v\|_1^2 \\
&\leq C \lambda_k \|(I - Q_k)v\|^2 + C \|v\|_1^2 \\
&\leq C \|v\|_1^2.
\end{aligned}
$$

Now we have proved (6.11) and (6.12). It remains to show that (6.11) and (6.12) imply (6.6) and (6.7) or A.7. This is done by the duality argument of Aubin and Nitsche as follows: We first note that (6.7) follows from (6.12) and the trivial inequality $\|Q_k v\| \leq \|v\|$ by Theorem B.4. To prove (6.6) we set $\varphi = A^{1-\alpha}(I - P_k)v \in M$. Then

$$(6.18) \qquad (A^{1-\alpha}(I - P_k)v, (I - P_k)v) = ((I - P_k)v, \varphi).$$

Let $w \in H^{1+\alpha}(\Omega) \cap H_0^1(\Omega)$ satisfy

$$A(w, \psi) = (\varphi, \psi), \quad \text{for all } \psi \in H_0^1(\Omega).$$

Choosing $\psi = (I - P_k)v$, we have

$$
\begin{aligned}
(6.19) \qquad ((I - P_k)v, \varphi) &= A(w, (I - P_k)v) = A(v, (I - P_k)w) \\
&\leq C \|v\|_1 \|(I - P_k)w\|_1 \\
&\leq C h_k^\alpha \|v\|_1 \|w\|_{1+\alpha} \\
&\leq C h_k^\alpha \|v\|_1 \|\varphi\|_{-1+\alpha}.
\end{aligned}
$$

85

Here we have used (6.11) and (6.4). Now for $\varphi \in M$

$$\|\varphi\|_{-1+\alpha} = \sup_{\theta \in H_0^1(\Omega)} \frac{(\varphi, \theta)}{\|\theta\|_{1-\alpha}} = \sup_{\theta \in H_0^1(\Omega)} \frac{(\varphi, Q_J \theta)}{\|\theta\|_{1-\alpha}}$$

$$\leq (A^{-(1-\alpha)}\varphi, \varphi)^{1/2} \sup_{\theta \in H_0^1(\Omega)} \frac{(A^{1-\alpha}Q_J\theta, Q_J\theta)^{1/2}}{\|\theta\|_{1-\alpha}}.$$

But

$$(A^{1-\alpha}Q_J\theta, Q_J\theta)^{1/2} \leq C\|\theta\|_{1-\alpha}$$

also follows, by Theorem B.4, from (6.12) and $\|Q_J\theta\| \leq \|\theta\|$. Thus for $\varphi \in M$

(6.20) $$\|\varphi\|_{-1+\alpha}^2 \leq C(A^{-(1-\alpha)}\varphi, \varphi).$$

Combining (6.18), (6.19) and (6.20) proves (6.6).

Thus we have proved that A.7 is satisfied and hence from Theorem 3.4 the multigrid V–cycle has a uniform reduction rate. In this example it is possible to show that A.1 and A.2 are satisfied as well so that Theorem 3.2 also yields the same result.

Equations with rough coefficients: The next example is that of the problem defined in (6.2) in which the coefficients are merely uniformly bounded and measurable. In such a case $A(\cdot, \cdot)^{1/2}$ defines a norm equivalent to the norm $\| \cdot \|_1$. We have already proved in the previous discussion that the conditions of Theorem 3.3 are satisfied. We can take the operators required for A.6 as $\overline{Q}_k = Q_k$, the L^2 projection operators. In the preceding development we proved (6.14) and (6.17) which are all that is needed for A.6. Thus we can invoke Theorem 3.3 to show that Algorithm I has an associated error operator whose contraction number deteriorates no worse than linearly in the number of levels.

We may also apply Theorem 3.3 to certain interface problems in which there may be large jumps in the coefficients. This situation was treated in [43]. By introducing appropriate weighted norms Condition A.6 was shown to hold and the multigrid convergence results are thus independent of the size of the jumps.

<u>Mesh refinement.</u> We also want to consider how the results of Section 3 apply to finite element approximations that utilize a locally refined mesh. Such mesh refinements are convenient for accurate modeling of problems with various types of singular behavior. As in [36], we consider for simplicity the case where Ω is a polygonal domain and the finite element space consists of piecewise linear functions, although we allow a very general form of refinement.

Following [45], we start with a coarse quasi-uniform triangulation $\mathcal{T}_1$. The refinement triangulation is defined in terms of a sequence of (open) mesh domains with

$$\Omega_J \subseteq \Omega_{J-1} \subseteq \ldots \subseteq \Omega_1 = \Omega.$$

The only restrictions on the mesh domains $\{\Omega_k\}$ are that the boundary of Ω_k, for $k > 1$, consists of edges of mesh triangles in the triangulation $\mathcal{T}_{k-1}$ and that there is at least one edge of $\mathcal{T}_{k-1}$ contained in Ω_k. These mesh domains control the region of refinement. If τ_{k-1}^i is a triangle contained in Ω_k, then it is broken into four smaller triangles (in the triangulation $\mathcal{T}_k$) by the lines connecting the midpoints of the edges. Alternatively, if τ_{k-1}^i is in the complement of Ω_k, then it is not subdivided but is directly included into the triangulation $\mathcal{T}_k$. A simple example of this construction is the case of a unit square with local refinement near the upper right hand corner. In this case we take $\Omega_k = \Omega$ for $k = 1, \ldots, j$ and $\Omega_k = (1 - 2^{j-k}, 1) \times (1 - 2^{j-k}, 1)$ for $k = j+1, \ldots, J$.

The space M_k is defined to be the set of piecewise linear functions with respect to the triangulation $\mathcal{T}_k$ which are continuous on Ω and vanish on $\partial\Omega$. The continuity condition implies that the finer grid nodes on a coarse-fine boundary are "slave nodes" in the sense that the values of the function are completely determined by the values of the function at the nearby coarse grid points. In this case, if $\Omega_{k-1} \neq \Omega_k$, then the subspace $\hat{M}_k$ on which we smooth is a proper nonzero subspace of M_k. In fact, we define $\hat{M}_k$ to be the functions in M_k which are zero outside of Ω_k. Thus we smooth on a given level just in the region where new nodes are being added in the refinement scheme. It is easy to see that the mesh corresponding to the space

87

$\hat{M}_k$ is quasi-uniform.

Set h_k to be the diameter of the smallest triangle in the k'th triangulation. By construction, $h_k \equiv 2^{1-k}h_1$.

It is shown in [34] that smoothers satisfying A.4a and A.5 are easily constructed. If we let $\overline{\overline{M}}_k$ be the space of piecewise linear functions on the fully refined mesh, then it is shown in [36] how to construct the operators required in A.8. Clearly A.7a holds here. Thus Theorem 3.5 may be applied to show the V–cycle yields a uniform reduction. Theorem 3.3 is also applicable in this case but the result is weaker.

Bibliographical Notes

The main applications of the multigrid methods are to second order elliptic problems (cf. [15] and [98]). The particular treatment given in this section is taken from [36] where for the first time a nested multigrid algorithm was formulated for the Dirichlet problem for domains with curved boundaries. In that paper the uniform convergence of the V–cycle with one smoothing step was proved using Theorem 3.2. Here we show that Theorem 3.4 is also applicable. The case of equations with rough coefficients is discussed in [43] where it is shown that Theorem 3.3 applies.

7. Perturbation of Forms

In this section we will continue to look at second order elliptic problems such as those considered in Section 6. In practice it is often inconvenient to compute the entries of the matrices coming from the quadratic form $A(\cdot,\cdot)$ of (6.1). It may be convenient to replace the form by a new form $\hat{A}(\cdot,\cdot)$ which is equivalent and, in a sense, close to $A(\cdot,\cdot)$. For example, for computational purposes, it may be necessary to replace the exact integration over a given triangle using a quadrature formula. To this end we consider a new approximate problem.

Find $u \in M$ such that

$$(7.1) \qquad \hat{A}(u,\chi) = F(\chi), \text{ for all } \chi \in M.$$

We will give examples of this later in this section.

We suppose that this new form is equivalent to the form $A(\cdot,\cdot)$. That is, there are positive constants C_0 and C_1 such that

$$(7.2) \qquad C_0 A(u,u) \le \hat{A}(u,u) \le C_1 A(u,u),$$

for all $u \in M$. We will also assume that the forms are close in the following sense:

A.13: For some number $\beta > 0$

$$|A(u,v) - \hat{A}(u,v)| \le Ch^\beta \|u\|_1 \|v\|_1,$$

for all u and $v \in M$.

Now we will see that we can check conditions A.1, A.2 and A.7 for $\hat{A}(\cdot,\cdot)$ by checking them for $A(\cdot,\cdot)$.

First, Lemma 3.1 says that, in view of (7.2), A.1 holds for $\hat{A}(\cdot,\cdot)$ if and only if it holds for $A(\cdot,\cdot)$. Now we can prove the following.

89

LEMMA 7.1. *Assume that (7.2) and A.13 hold. Then A.2 holds for $A(\cdot,\cdot)$ if and only if it holds for $\hat{A}(\cdot,\cdot)$.*

Proof: Clearly λ_k is comparable in size to $\hat{\lambda}_k$, the largest eigenvalue of $\hat{A}_k$, and $\lambda_k \approx h_k^{-2}$. Now

$$(7.3) \qquad \hat{\lambda}_k^{-1}\|\hat{A}_k\hat{P}_k w\|^2 = \hat{\lambda}_k^{-1} \sup_{\varphi\in M_k}\left(\frac{\hat{A}(w,\varphi)}{\|\varphi\|}\right)^2.$$

Clearly,

$$(7.4) \qquad \sup_{\varphi\in M_k}\frac{\hat{A}(w,\varphi)}{\|\varphi\|} \le \sup_{\varphi\in M_k}\frac{A(w,\varphi)}{\|\varphi\|} + \sup_{\varphi\in M_k}\frac{|A(w,\varphi)-\hat{A}(w,\varphi)|}{\|\varphi\|}.$$

Now suppose A.2 holds for $A(\cdot,\cdot)$. Then, for $w \in M_\ell$, $\ell \le k$,

$$(7.5) \qquad \hat{\lambda}_k^{-1}\left(\frac{A(w,\varphi)}{\|\varphi\|}\right)^2 \le \tilde{C}\varepsilon^{2(k-\ell)}\hat{A}(w,w)$$

for some $0 < \varepsilon < 1$. Also by A.13

$$(7.6) \qquad \sup_{\varphi\in M_k}\frac{|A(w,\varphi)-\hat{A}(w,\varphi)|}{\|\varphi\|} \le Ch^\beta\|w\|_1 \sup_{\varphi\in M_k}\frac{A(\varphi,\varphi)^{1/2}}{\|\varphi\|}$$
$$\le C\hat{\lambda}_k h^\beta \hat{A}(w,w).$$

Combining (7.3)–(7.6) proves A.2 for $\hat{A}(\cdot,\cdot)$. The lemma now follows by symmetry.

A similar result holds for condition A.7. We have the following:

LEMMA 7.2. *Assume that (7.2) holds and that A.13 holds with $\beta = \alpha$. Then A.7 holds for $A(\cdot,\cdot)$ if and only if it holds for $\hat{A}(\cdot,\cdot)$.*

Proof: Assume that A.7 holds for $A(\cdot,\cdot)$. Clearly, by (7.2) and Theorem B.4, the first inequality also holds for $\hat{A}(\cdot,\cdot)$. Now, again using (7.2) and Theorem B.4,

$$(7.7)$$
$$(\hat{A}^{1-\alpha}(I-\hat{P}_k)u,(I-\hat{P}_k)u)^{1/2} \le C((A^{1-\alpha}(I-P_k)u,(I-P_k)u)^{1/2}$$
$$+ \|(P_k-\hat{P}_k)u\|_1)$$
$$\le C(h_k^\alpha\|u\|_1 + \|(P_k-\hat{P}_k)u\|_1).$$

Set $v = (P_k - \hat{P}_k)u$. Then

(7.8) $\qquad \|(P_k - \hat{P}_k)u\|_1^2 \le CA((P_k - \hat{P}_k)u, v)$

$$= C(A(u,v) - \hat{A}(u,v) + \hat{A}(\hat{P}_k u, v) - A(\hat{P}_k u, v))$$

$$\le Ch^\alpha \|u\|_1 \|v\|_1,$$

by A.13. Clearly (7.8) implies that

$$\|(P_k - \hat{P}_k)u\|_1 \le Ch^\alpha \|u\|_1.$$

Hence (7.7) and (7.8) imply the second inequality in A.7. The lemma now follows by symmetry.

Applications

1. Finite elements with quadrature.

We shall consider the form (6.1) with $a = 0$, a_{ij} smooth and $\Omega \subset R^2$ a polygonal domain. We define on the nested set of quasi–uniform triangulations $\mathcal{T}_k$ of Ω the nested set of finite element spaces M_k, $k = 1, \ldots, J$ consisting of polynomials on each triangle of degree less than or equal to K, and continuous on Ω. Now the usual way to define a quadrature scheme is as follows. Let $\tau_J^i \in \mathcal{T}_J$ and let τ be a (closed) reference triangle. On τ we introduce the reference quadrature

$$\int_\tau \hat{\phi}(\hat{x}) d\hat{x} \approx \sum_{\ell=1}^L w_\ell \hat{\phi}(b_\ell)$$

where w_ℓ are positive weights and $b_\ell \in \tau$ are the quadrature points. On each fine grid triangle τ_J^i we take the quadrature rule to be

$$\int_{\tau_J^i} \phi(x) dx \approx \sum_{\ell=1}^L w_{J,\ell}^i \phi(b_{J,\ell}^i) \equiv Q_J^i[\phi],$$

where $\phi(x) = \hat{\phi}(\hat{x})$ and the weights $w_{J,\ell}^i$ and points $b_{J,\ell}^i$ are defined in terms of w_ℓ and b_ℓ by means of an affine mapping from τ_J^i onto τ which takes each x in τ_J^i into $\hat{x}$ in τ.

91

The condition which we need on $Q^i_J[\phi]$ is that

(7.9) $$Q^i_J[\phi] = \int_{r^i_j} \phi(x)dx$$

if ϕ is a polynomial of degree $2K - 2$. The approximate form $\hat{A}$ is now defined by

$$\hat{A}(v, w) = \sum_{r^i_j \in T_J} Q^i_J \Big[\sum_{i,j=1}^{2} \Big(a_{ij} \frac{\partial v}{\partial x_i} \frac{\partial w}{\partial x_j} \Big) \Big],$$

for all $v, w \in M_J$.

It can be shown that with (7.9), (7.2) holds (cf. [65]). Furthermore, the exactness condition also implies A.13. Thus we may invoke Theorem 3.2 to conclude that the V–cycle has a uniform rate of reduction provided we use an appropriate smoother.

2. Finite difference example.

We consider, for simplicity, a domain $\Omega \subset R^2$ made up of the union of squares of side $h = h_J$. We define the mesh domain

$$\overline{\Omega}_h = \{x \in \overline{\Omega} \mid x = (\ell h, mh), \text{ for } \ell, m \text{ integers}\}.$$

Set $\Omega_h = \overline{\Omega}_h \cap \Omega$ and $\partial \Omega_h = \overline{\Omega}_h \cap \partial \Omega$. Let S_h denote the set of grid functions on $\overline{\Omega}_h$ which vanish on $\partial \Omega_h$. For $V \in S_h$ set $V_{i,j} = V(ih, jh)$.

We shall consider the special case of (6.1) with

(7.10) $$Lu = -\frac{\partial}{\partial x_1} a \frac{\partial u}{\partial x_1} - \frac{\partial}{\partial x_2} b \frac{\partial u}{\partial x_2} + cu,$$

where a and b are positive functions and $c \geq 0$. The following is a standard five-point finite difference approximation of (6.2). For $F \in S_h$, find $U \in S_h$ satisfying

$$L_h U = F,$$

where $L_h : S_h \to S_h$ is given by

(7.11) $$(L_h V)_{i,j} = a_{i+1/2,j}(V_{i,j} - V_{i+1,j}) + a_{i-1/2,j}(V_{i,j} - V_{i-1,j})$$
$$+ b_{i,j+1/2}(V_{i,j} - V_{i,j+1}) + b_{i,j-1/2}(V_{i,j} - V_{i,j-1})$$
$$+ h^2 c_{i,j} V_{i,j}.$$

92

Here $f_{s,t} = f(sh, th)$ for any function f.

Now we can relate the above scheme to a certain finite element scheme by considering it as coming from certain quadratures. We do this as follows:

Consider the triangulation $\mathcal{T}_J$ which results from dividing each square on the finite difference grid into two triangles with positively sloping diagonals. Let M_J be the space of continuous piecewise linear functions on $\mathcal{T}_J$. Assume further that $\mathcal{T}_J$ results from successive subdivisions of an original coarse triangulation $\mathcal{T}_1$.

Let τ_j^i be a typical triangle with

1. m_ℓ^i, the midpoint of the edge parallel to the x_ℓ axis, $\ell = 1, 2$,

and

2. $\tilde{v}_\ell^i$, $\ell = 1, 2, 3$, its vertices.

We use the following three quadrature formulae, respectively, for the three terms coming from (7.10):

$$\int_{\tau_j^i} \phi \, dx \approx \frac{h^2}{2} \phi(m_1^i)$$

$$\int_{\tau_j^i} \phi \, dx \approx \frac{h^2}{2} \phi(m_2^i)$$

and

$$\int_{\tau_j^i} \phi \, dx \approx \frac{h^2}{6} [\phi(\tilde{v}_1^i) + \phi(\tilde{v}_2^i) + \phi(\tilde{v}_3^i)].$$

Let $\hat{A}(\cdot, \cdot)$ be the resulting quadratic form on the space M_J. It can be verified that the stiffness matrix associated with $\hat{A}(\cdot, \cdot)$ is equal to the matrix of the finite difference scheme (7.11). It is also easy to verify that (7.2) holds, where $A(\cdot, \cdot)$ is the quadratic form associated with (7.10). Furthermore, the above quadrature schemes are exact on constants. Consequently, since $K = 1$ ($2K - 2 = 0$), it follows easily that A.13 holds, if a, b and c are Hölder continuous functions with exponent β. Thus with the prolongation coming from the finite element embedding and using, e.g., Gauss–Seidel as smoother, Theorem 3.2 implies that the multigrid algorithm gives rise to a uniform reduction rate.

It is easy to check that the stiffness matrix corresponding to $\hat{A}$ associated with any level k comes from a five point difference operator. The coefficients may be computed recursively from the coefficients on the finest level.

Bibliographical Notes

A discussion of the use of quadrature in finite element methods may be found in [65] and multigrid results for finite elements with quadrature were given in [91]. It was assumed there that coarse grid quadrature formulae were used on the coarser levels and that full elliptic regularity (i.e., A.7 with $\alpha = 1$) was satisfied. This means that Ω must be convex and the coefficients must be smooth. The present method and analysis are in [25].

The treatment of finite difference applications may be found in [98]. It is not clear that the theorems there are applicable to our example. The formulation and analysis given here comes from [25].

8. Pseudodifferential operators of order minus one

Weakly singular integral operators arise, for example, in the approximation of the solution of elliptic boundary value problems by the so-called boundary element method. In this section we shall consider pseudodifferential operators of order minus one which give rise to coercive quadratic forms on $H^{-1/2}(\Omega)$ (to be defined). The aforementioned integral operators will be our main example. We shall develop multigrid algorithms for solving the associated discrete equations.

For purposes of exposition we shall consider Ω to be a plane polygonal domain which we may think of as being embedded in R^3.

For $s \geq 0$, an integer, we denote by $H^s(\Omega)$ the usual Sobolev space with norm given by

$$\|u\|_s^2 = \sum_{|\alpha| \leq s} \int_\Omega |D^\alpha u|^2 dx.$$

For $s > 0$, not an integer, $H^s(\Omega)$ is defined by interpolation between spaces defined for successive integers. The space $H^{-1}(\Omega)$ is defined as the completion of $L^2(\Omega)$ with respect to the norm defined by

$$\|u\|_{-1} = \sup_{\varphi \in H^1(\Omega)} \frac{(u, \varphi)}{\|\varphi\|_1}.$$

For $0 < s < 1$, $H^{-s}(\Omega)$ is defined by interpolation between $L^2(\Omega)$ and $H^{-1}(\Omega)$. These spaces are Hilbert spaces with inner products denoted by $< \cdot, \cdot >_{-s}$ and $< \cdot, \cdot >_0 = (\cdot, \cdot)$, the $L^2(\Omega)$ inner product. We now consider a positive definite bilinear form $\mathcal{V}(\cdot, \cdot)$ on $H^{-1/2}(\Omega) \times H^{-1/2}(\Omega)$ and assume that for positive constants C_0 and C_1

$$C_0 \|v\|_{-1/2}^2 \leq \mathcal{V}(v, v) \leq C_1 \|v\|_{-1/2}^2,$$

for all $v \in H^{-1/2}(\Omega)$. Let F be a bounded linear functional on $H^{-1/2}(\Omega)$ and consider the problem: Find $U \in H^{-1/2}(\Omega)$ such that

$$\mathcal{V}(U, \theta) = F(\theta), \text{ for all } \theta \in H^{-1/2}(\Omega).$$

By the Riesz Representation Theorem this problem has a unique solution. Our main example of such a form is given by

(8.1)
$$\mathcal{V}(u,v) = \int_\Omega \int_\Omega \frac{u(s_1)v(s_2)}{|s_1 - s_2|} ds_1 ds_2.$$

We will consider this example in detail later in this section, but for now we do not specialize to this form.

The approximate problem

As before we consider the spaces M_k with

$$M_1 \subset M_2 \subset \cdots \subset M_J = M \subset H^{-1/2}(\Omega).$$

The approximate problem in M is: Find $u \in M$ such that

(8.2)
$$\mathcal{V}(u, \varphi) = F(\varphi), \text{ for all } \varphi \in M.$$

In order to motivate and simplify our discussion assume that $M \subset H^1(\Omega)$. This is convenient but not necessary. Let $\mathcal{V}_J : M \to M$ be defined by

$$(\mathcal{V}_J u, \varphi) = \mathcal{V}(u, \varphi), \text{ for all } \varphi \in M.$$

Now the extreme eigenvalues, $\mu_{\min}$ and $\mu_{\max}$, of $\mathcal{V}_J \varphi_i = \mu_i \varphi_i$ are the extremes of the Rayleigh quotient

$$\mu_{\min} \leq \frac{\mathcal{V}(v,v)}{(v,v)} \leq \mu_{\max}, \quad 0 \neq v \in M,$$

or

$$\mu_{\max}^{-1} \leq \frac{(v,v)}{\mathcal{V}(v,v)} \leq \mu_{\min}^{-1}.$$

This behaves like

(8.3)
$$\frac{(v,v)}{\mathcal{V}(v,v)} \approx \frac{(v,v)}{\|v\|_{-1/2}^2} \approx \frac{(A_J^{1/2}v,v)_J}{\|v\|^2}$$

where A_J is analogous to a second order elliptic operator defined on M by

$$(A_J w, \chi)_J = \int_\Omega (\nabla w \cdot \nabla \chi + w\chi)dx,$$

for all w and $\chi \in M$. Here $(\cdot, \cdot)_J$ is an appropriately defined inner product, comparable to the $L^2(\Omega)$ inner product.

Suppose further that M consists of continuous, piecewise linear functions on the triangulation as in Section 2. Then, following the argument in Section 2, noting that the eigenvalues of $A_J^{1/2}$ are the square roots of those of A_J, we have

(8.4)
$$C_0 \leq \frac{(A_J^{1/2} v, v)_J}{\|v\|^2} \leq C_1 h_J^{-1}.$$

Because of (8.3) and (8.4)

$$\mu_{\min} \geq C h_J \text{ and } \mu_{\max} \leq C.$$

In order to see what the difficulty is in the present setting we consider for a moment the special case (8.1). The operator $\mathcal{V} : H^{-1/2}(\Omega) \to H^{1/2}(\Omega)$ defined by

$$(\mathcal{V}u)(s) = \int_\Omega \frac{u(s_1)ds_1}{|s_1 - s|}$$

gives rise to the form (8.1) and is compact (as an operator from $H^{-1/2}(\Omega)$ to $H^{-1/2}(\Omega)$) because of the compact embedding of $H^{1/2}(\Omega)$ in $H^{-1/2}(\Omega)$ (cf [131]). The discrete operator $\hat{\mathcal{V}}_J : M \to M$ defined by

$$(\hat{\mathcal{V}}_J w, \chi) = \mathcal{V}(w, \chi), \text{ for all } w \text{ and } \chi \in M$$

is SPD and has eigenvalues μ_i (with corresponding orthonormal eigenvectors φ_i) which tend to zero with h_J. The highly oscillatory eigenvectors correspond to small eigenvalues so that a simple linear iteration such as

$$u^{n+1} = u^n - \tau \hat{\mathcal{V}}_J(u^n - u)$$

is not convergent for any choice of τ which damps the highly oscillatory components of the error, since such a τ would necessarily be large and hence would amplify the smooth components . Thus the naive multigrid approach will fail.

The remedy for this problem in the general case lies in properly defining the discrete operators $\mathcal{V}_k$. Define $\mathcal{V}_k : M_k \to M_k$ by

$$< \mathcal{V}_k w, \chi >_{-1} = \mathcal{V}(w, \chi),$$

for all $w, \chi \in M_k$. Then $\mathcal{V}_k$ is SPD with respect to $< \cdot, \cdot >_{-1}$. Its eigenvalues $\{\nu_i\}$ are positive and the corresponding eigenfunctions $\{\psi_i\}$ can be chosen to be orthonormal in $< \cdot, \cdot >_{-1}$. The extremal eigenvalues are the extremes of the Rayleigh quotient

$$\nu_{min} \leq \frac{\mathcal{V}(v, v)}{\|v\|_{-1}^2} \leq \nu_{max}.$$

The denominator is weaker than the numerator and hence $\nu_{min} \approx 1$ and $\nu_{max} \approx h_J^{-1}$.

Consider now the linear iteration

$$v^{n+1} = v^n - \beta \mathcal{V}_J(v^n - v).$$

If

$$v - v^0 = \sum_{i=1}^{N_J} a_i \psi_i,$$

then

$$v - v^n = \sum_{i=1}^{N_J} a_i(1 - \beta \nu_i)^n \psi_i.$$

Hence for $\beta \approx \nu_J^{-1}$ and $\beta \geq C\nu_J^{-1}$ we see that, since $\{\nu_i\}$ is increasing, the high frequencies of the error are reduced leaving the low frequencies relatively unchanged. This is the situation in which we expect a multilevel approach to be effective.

In order to define a multigrid algorithm it remains to define smoothers $R_k : M_k \to M_k$. The smoothers defined in Section 5 do not seem to be adaptable to the theory in this section. We will need the orthogonal projector $\mathcal{P}_k : H^{-1}(\Omega) \to M_k$ defined by

$$< \mathcal{P}_k w, \varphi >_{-1} = < w, \varphi >_{-1}, \text{ for all } \varphi \in M_k.$$

To define the smoother we introduce on M_k an inner product $[\cdot, \cdot]_k$. This inner product will be defined a little later. Care must be taken in its choice in order that

98

there are no difficulties in the implementation of the multigrid algorithm. In terms of $[\cdot,\cdot]_k$ we define $R_k : M_k \to M_k$ as

$$
(8.5) \qquad\qquad [R_k w, \theta]_k = \frac{1}{\lambda_k} < w, \theta >_{-1}
$$

with

$$
\lambda_k \leq \overline{\lambda}_k \text{ and } \lambda_k = \sup_{\theta \in M_k} \frac{\mathcal{V}(\theta, \theta)}{[\theta, \theta]_k}.
$$

We also assume that $\overline{\lambda}_k \leq C\lambda_k$. Now, with this definition we have

$$
\begin{aligned}
\mathcal{V}(R_k \mathcal{V}_k \varphi, \varphi) &\leq \mathcal{V}(R_k \mathcal{V}_k \varphi, R_k \mathcal{V}_k \varphi)^{1/2} \mathcal{V}^{1/2}(\varphi, \varphi) \\
&\leq \overline{\lambda}_k^{1/2} [R_k \mathcal{V}_k \varphi, R_k \mathcal{V}_k \varphi]_k^{1/2} \mathcal{V}^{1/2}(\varphi, \varphi) \\
&= < \mathcal{V}_k \varphi, R_k \mathcal{V}_k \varphi >_{-1}^{1/2} \mathcal{V}^{1/2}(\varphi, \varphi) \\
&= \mathcal{V}(R_k \mathcal{V}_k \varphi, \varphi)^{1/2} \mathcal{V}^{1/2}(\varphi, \varphi).
\end{aligned}
$$

Hence

$$
\mathcal{V}(R_k \mathcal{V}_k \varphi, \varphi) \leq \mathcal{V}(\varphi, \varphi).
$$

Now A.5, with $\theta = 1$, follows.

Noting that R_k is SPD, we may use R_k in Algorithm I which is the following:

Algorithm I in this case:

0) $B_1 = \mathcal{V}_1^{-1}$

For $k > 1$, B_k is defined in terms of B_{k-1} as follows. Let $g \in M_k$.

1) $x^1 = R_k g$

2) $x^2 = x^1 - q$, where $q = B_{k-1} \mathcal{P}_{k-1}(\mathcal{V}_k x^1 - g)$

3) $B_k g = x^2 - R_k(\mathcal{V}_k x^2 - g)$.

We will see that with R_k appropriately defined, the implementation of this algorithm does not involve the inner product $< \cdot, \cdot >_{-1}$.

As stated earlier, for the purpose of exposition and analysis, we shall assume that M_1 is the usual coarse space of piecewise linear functions on Ω. We define M_k as before. No boundary conditions are imposed. Define now the discrete inner product

$(\cdot, \cdot)_k$ on M_k by

$$(v, v)_k = \frac{1}{3} \sum_{\tau_k^i \in \mathcal{T}_k} |\tau_k^i| [v(x_k^{i,1})^2 + v(x_k^{i,2})^2 + v(x_k^{i,3})^2].$$

Here $|\tau_k^i|$ denotes the area of τ_k^i and $x_k^{i,\ell}$, $\ell = 1, 2, 3$, denote the vertices of τ_k^i. It is known that

(8.6)
$$|(v, w)_k - (v, w)| \le C h_k \|v\| \, \|w\|_1,$$

for all $v \in M_k$ and $w \in M$. Also it follows that there are positive constants c and C, independent of k, such that

$$c(v, v)_k \le (v, v) \le C(v, v)_k, \text{ for all } v \in M_k.$$

Now let

$$A(v, w) = \int_\Omega (\nabla v \cdot \nabla w + vw) dx$$

and $\|u\|_1 = A(u, u)^{1/2}$. As before we define $A_k : M_k \to M_k$ by

$$(A_k w, \varphi)_k = A(w, \varphi), \text{ for all } w, \varphi \in M_k.$$

We now define on M_k

(8.7)
$$[u, v]_k \equiv (A_k^{-1} u, v)_k.$$

We will see that with this definition the multigrid algorithm involves only the action of A_k (not A_k^{-1}). The next lemma is central.

LEMMA 8.1. *There exist positive constants C_0 and C_1 which are independent of J such that*

$$C_0 \|v\|_{-1}^2 \le [v, v]_k \le C_1 \|v\|_{-1}^2.$$

Proof: Now $M_k \subset H^1(\Omega)$ so that, for $v \in H^1(\Omega)$,

(8.8)
$$[v, v]_k = (A_k^{-1} v, v)_k = \sup_{\varphi \in M_k} \frac{(v, A_k^{-1/2} \varphi)_k^2}{(\varphi, \varphi)_k} = \sup_{\theta \in M_k} \frac{(v, \theta)_k^2}{A(\theta, \theta)}.$$

100

Using (8.6), we have

(8.9) $\qquad |(v,\theta)_k| \leq |(v,\theta)_k - (v,\theta)| + |(v,\theta)| \leq Ch_k\|v\|\,\|\theta\|_1 + \|v\|_{-1}\|\theta\|_1.$

Also $A(\varphi,\varphi) \leq Ch_k^{-2}\|\varphi\|^2$ for $\varphi \in M_k$. Hence for $v \in M_k$

(8.10) $\qquad \|v\|^2 = \sup_{\varphi \in M_k} \frac{(v,\varphi)^2}{\|\varphi\|^2} \leq Ch_k^{-2} \sup_{\varphi \in M_k} \frac{(v,\varphi)^2}{A(\varphi,\varphi)} \leq Ch_k^{-2}\|v\|_{-1}^2.$

Similarly

(8.11) $\qquad\qquad\qquad \|v\|^2 \leq Ch_k^{-2} \sup_{\varphi \in M_k} \frac{(v,\varphi)_k^2}{A(\varphi,\varphi)}.$

Hence, using (8.9) and (8.10), we see that

$$|(v,\theta)_k| \leq C\|v\|_{-1}\|\theta\|_1$$

and therefore

$$[v,v]_k \leq C\|v\|_{-1}^2,$$

which is the upper inequality of Lemma 8.1. We next prove that $\|v\|_{-1}^2 \leq C[v,v]_k$. To this end let $Q_k : L^2(\Omega) \to M_k$ be the $L^2(\Omega)$ orthogonal projection onto M_k. It is known (cf. [50]) that for $\varphi \in H^1(\Omega)$

(8.12) $\qquad\qquad\qquad\qquad \|Q_k\varphi\|_1 \leq C\|\varphi\|_1.$

For $v \in M_k$

(8.13) $\qquad \|v\|_{-1} = \sup_{\varphi \in H^1(\Omega)} \frac{(v,\varphi)}{\|\varphi\|_1} = \sup_{\varphi \in H^1(\Omega)} \frac{(v,Q_k\varphi)}{\|\varphi\|_1}$

$\qquad\qquad\qquad = \sup_{\varphi \in H^1(\Omega)} \frac{(v,Q_k\varphi)}{\|Q_k\varphi\|_1} \frac{\|Q_k\varphi\|_1}{\|\varphi\|_1}$

$\qquad\qquad\qquad \leq C \sup_{\psi \in M_k} \frac{(v,\psi)}{\|\psi\|_1}.$

Again we use (8.6) to obtain

$$|(v,\theta)| \leq Ch_k\|v\|\,\|\theta\|_1 + |(v,\theta)_k|,$$

or

$$(8.14) \qquad \frac{|(v,\theta)|}{\|\theta\|_1} \le Ch_k\|v\| + \frac{|(v,\theta)_k|}{\|\theta\|_1}.$$

The result now follows from (8.13), (8.14), (8.11) and (8.8).

Multigrid analysis

We shall first consider the general case. Since we cannot show that A.2 or A.7 is satisfied, we want to apply the weaker result Theorem 3.3. To apply Theorem 3.3 we let Q_k be the $L^2(\Omega)$ orthogonal projection and show that

$$(8.15) \qquad \|(Q_k - Q_{k-1})u\|_{-1}^2 \le C\lambda_k^{-1}\mathcal{V}(u,u)$$

and

$$(8.16) \qquad \mathcal{V}(Q_ku, Q_ku) \le C\mathcal{V}(u,u)$$

are satisfied. To prove (8.15) we first note that

$$\|(I - Q_k)u\|_{-1} = \sup_{v \in H^1(\Omega)} \frac{((I-Q_k)u, v)}{\|v\|_1} \le \sup_{v \in H^1(\Omega)} \frac{\|u\| \, \|(I-Q_k)v\|}{\|v\|_1} \le Ch_k\|u\|.$$

Here we used the well known approximation property

$$\|(I - Q_k)v\| \le Ch_k\|v\|_1,$$

for all $v \in H^1(\Omega)$. Since $M_k \subset H^1(\Omega)$, Q_k is well-defined as a map from $H^{-1}(\Omega)$ to $L^2(\Omega)$. Hence by (8.12), with $u \in H^{-1}(\Omega)$

$$\|Q_ku\|_{-1} = \sup_{\varphi \in H^1(\Omega)} \frac{(Q_ku, \varphi)}{\|\varphi\|_1}$$

$$= \sup_{\varphi \in H^1(\Omega)} \frac{(u, Q_k\varphi)}{\|\varphi\|_1}$$

$$\le C\|u\|_{-1}.$$

Therefore by interpolation

$$\|(I - Q_k)u\|_{-1}^2 \le Ch_k\|u\|_{-1/2}^2 \le Ch_k\mathcal{V}(u,u).$$

This proves (8.15) once we show that

$$h_k \le C\lambda_k^{-1},$$

where

$$\lambda_k = \sup_{\varphi \in M_k} \frac{\mathcal{V}(\varphi, \varphi)}{[\varphi, \varphi]_k} \le \sup_{\varphi \in M_k} \frac{\|\varphi\|_{-1/2}^2}{\|\varphi\|_{-1}^2}.$$

But, by Theorem B.1 of Appendix B and (8.10),

$$\|\varphi\|_{-1/2}^2 \le \|\varphi\| \, \|\varphi\|_{-1} \le C h_k^{-1} \|\varphi\|_{-1}^2.$$

Hence $\lambda_k \le C h_k^{-1}$. In a similar way we can prove (8.16).

Now we want to check the lower inequality of A.3. But this is clear since

$$\|w\|_{-1}^2 = \overline{\lambda}_k [R_k w, w]_k = \overline{\lambda}_k [R_k^{1/2} w, R_k^{1/2} w]_k$$
$$\le C\lambda_k < R_k^{1/2} w, R_k^{1/2} w >_{-1}$$
$$= C\lambda_k < R_k w, w >_{-1}$$

since R_k is symmetric in both $[\cdot, \cdot]_k$ and $< \cdot, \cdot >_{-1}$ and positive definite. Now it is easy to see that if R_k is symmetric, then the lower inequality in A.3 implies A.4. Hence Theorem 3.3 may be applied with $\hat{M}_k \equiv M_k$. We conclude that

$$\mathcal{V}((I - B_J \mathcal{V}_J) u, u) \le \left(1 - \frac{1}{CJ}\right) \mathcal{V}(u, u)$$

where B_J is defined with R_k given by (8.5).

We now consider the particular case of the form (8.1). Now we know more about the form $\mathcal{V}(\cdot, \cdot)$ and hence are able to prove more. Recall that

$$\mathcal{V}(u, v) = \int_\Omega \int_\Omega \frac{u(s_1) v(s_2)}{|s_1 - s_2|} ds_1 ds_2$$

where Ω is a polygonal domain in R^2. Let S be a smooth closed surface in R^3 (a two dimensional manifold) with $\Omega \subset S$. Set

$$\mathcal{V}_{(S)}(u, v) = \int_S \int_S \frac{u(s_1) v(s_2)}{|s_1 - s_2|} ds_1 ds_2.$$

103

It is shown by [131] that

$$C_0\|v\|^2_{-1/2,S} \leq \mathcal{V}_{(S)}(v,v) \leq C_1\|v\|^2_{-1/2,S}.$$

The operator

$$(\mathcal{V}_{(S)}u)(s_2) = \int_S \frac{u(s_1)}{|s_1-s_2|}ds_1$$

is an isomorphism of $H^s(S)$ onto $H^{s+1}(S)$ for all real s. That is, for each s,

(8.17) $$\qquad C_0\|w\|^2_{s,S} \leq \|\mathcal{V}_{(S)}w\|^2_{s+1,S} \leq C_1\|w\|^2_{s,S},$$

for all $w \in H^s(S)$.

We need the following lemma.

LEMMA 8.2. *Let* $\sigma \in L^2(\Omega)$ *and* $\tilde{\sigma}$ *denote the extension by zero of* σ *to* $L^2(S)$. *There exist positive constants* C_2 *and* C_3 *such that for* $s \in [-1,0]$

$$C_2\|\sigma\|^2_s \leq \|\tilde{\sigma}\|^2_{s,S} \leq C_3\|\sigma\|^2_s.$$

We shall delay the proof of this lemma until later.

Now, by Lemma 8.2

$$\mathcal{V}(\sigma,\sigma) = \mathcal{V}_{(S)}(\tilde{\sigma},\tilde{\sigma}) \approx \|\tilde{\sigma}\|^2_{-1/2,S} \approx \|\sigma\|^2_{-1/2}.$$

In order to apply Theorem 3.2 we first prove that A.2 is satisfied, i.e. for some $0 < \epsilon < 1$,

(8.18) $$\qquad \lambda_k^{-1}\|\mathcal{V}_k w\|^2_{-1} \leq (\tilde{C}\epsilon^{k-\ell})^2 \mathcal{V}(w,w),$$

for all $w \in M_\ell$ with $\ell \leq k$. Now

(8.19) $$\qquad \begin{aligned} \|\mathcal{V}_k w\|^2_{-1} &= <\mathcal{V}_k w, \mathcal{V}_k w>_{-1} = \mathcal{V}(w,\mathcal{V}_k w) \\ &= (\mathcal{V}_{(S)}\tilde{w}, \widetilde{\mathcal{V}_k w})_S \\ &\leq \|\mathcal{V}_{(S)}\tilde{w}\|_{1,S}\|\widetilde{\mathcal{V}_k w}\|_{-1,S} \\ &\leq C\|w\| \, \|\mathcal{V}_k w\|_{-1}. \end{aligned}$$

In the last inequality we used (8.17) and Lemma 8.2. Hence

$$\|\mathcal{V}_k w\|_{-1}^2 \leq C \|w\|^2.$$

Now

$$\|w\|^2 \leq C h_\ell^{-1} \|w\|_{-1/2}^2, \quad w \in M_\ell.$$

This follows from (8.10) and interpolation. Hence

$$\lambda_k^{-1} \|\mathcal{V}_k w\|_{-1}^2 \leq C (\lambda_k h_\ell)^{-1} \|w\|_{-1/2}^2 \leq C (\lambda_k h_\ell)^{-1} \mathcal{V}(w, w).$$

We finally need that

$$\lambda_k^{-1} \leq C h_k.$$

Now

$$(w, w)_k^{1/2} \leq C \|w\|$$

and

$$(A_k^{-1} w, w)_k^{1/2} \leq C \|w\|_{-1}.$$

By interpolation it follows that

$$(A_k^{-1/2} w, w)_k^{1/2} \leq C \|w\|_{-1/2}.$$

Now

$$\lambda_k = \sup_{\theta \in M_h} \frac{\mathcal{V}(\theta, \theta)}{[\theta, \theta]_k} \geq C \sup_{\theta \in M_h} \frac{\|\theta\|_{-1/2}^2}{[\theta, \theta]_k} \geq C \sup_{\theta \in M_h} \frac{(A_k^{-1/2} \theta, \theta)_k}{(A_k^{-1} \theta, \theta)_k}$$

$$= C \sup_{\theta \in M_h} \frac{(A_k^{1/2} \theta, \theta)}{(\theta, \theta)_k} \geq C h_k^{-1}.$$

Thus

$$\lambda_k^{-1} \|\mathcal{V}_k w\|_{-1}^2 \leq C (h_k/h_\ell) \mathcal{V}(w, w).$$

Since $h_k/h_\ell = \left(\frac{1}{2}\right)^{(k-\ell)}$, A.2 is satisfied with $\epsilon = 1/\sqrt{2}$.

The final ingredient in order to prove a uniform multigrid estimate (i.e., apply Theorem 3.2) is A.1, which in this case is the estimate

$$(8.20) \qquad \mathcal{V}(w, w) \leq C_0 \left[\mathcal{V}(P_1 w, w) + \sum_{k=2}^{J} \lambda_k^{-1} \|\mathcal{V}_k P_k w\|_{-1}^2 \right].$$

By Lemma 3.1 it suffices to show this for the equivalent form

$$\check{\mathcal{V}}(u, v) = < u, v >_{-1/2} .$$

For this purpose define $\check{\mathcal{V}}_k$ and $\check{P}_k$ by

$$< \check{\mathcal{V}}_k v, \theta >_{-1} = < v, \theta >_{-1/2}, \quad v, \theta \in M_k$$

and

$$< \check{P}_k w, \theta >_{-1/2} = < w, \theta >_{-1/2}, \quad \theta \in M_k.$$

It suffices to prove "full regularity" for this operator. That is

$$(8.21) \qquad < (I - \check{P}_{k-1})v, v >_{-1/2} \leq C \lambda_k^{-1} \|\check{\mathcal{V}}_k v\|_{-1}^2,$$

for all $v \in M_k$. Taking $v = \check{P}_k w$ in (8.21) (with $\check{P}_0 = 0$) and summing from 1 to J yields (8.20) for the equivalent operator $\check{\mathcal{V}}$. This proves A.1 once (8.21) is known. The proof of (8.21) may be found in [29]. Hence all of the conditions of Theorem 3.2 are satisfied and we conclude that the V–cycle with one smoothing yields a uniform reduction rate.

<u>Remark</u>: We took $M \subset H^1(\Omega)$ for convenience. There is no fundamental reason why M need consist of continuous piecewise polynomials. In fact, if M is made up of piecewise constant functions then the approximation (8.2) is well defined. An example is given in [29] in which $M_1 \subset M_2 \subset \cdots \subset M_J \equiv M$ consist of piecewise constants. The operator R_k is a little harder to define and the proof of Lemma 8.1 a bit more technical.

<u>Remark</u>: The choice of the operator R_k is crucial. Since it and the operator $\mathcal{V}_k$ both involve the computationally inconvenient form $< \cdot, \cdot >_{-1}$, it turns out that

this form never enters explicitly into the computation. Only the action of A_k is required. This involves only the stiffness matrix which is sparse.

Proof of Lemma 8.2: It is obvious for $s = 0$. Furthermore

$$\|\tilde{\sigma}\|_{-1} = \sup_{\varphi \in H^1(S)} \frac{(\tilde{\sigma}, \varphi)_S}{\|\varphi\|_{1,S}} \leq \sup_{\varphi \in H^1(S)} \frac{(\sigma, \varphi)}{\|\varphi\|_1} \leq \|\sigma\|_{-1}.$$

The upper inequality follows by interpolation. To prove the lower inequality let $E : H^s(\Omega) \to H^s(S)$ for $s \in [0,1]$ be an extension operator which is bounded, uniformly in s (the existence of such an extension is well known, cf. [96]). Then

$$\|\sigma\|_{-s} = \sup_{\varphi \in H^s(\Omega)} \frac{(\sigma, \varphi)}{\|\varphi\|_s} \leq C \sup_{\varphi \in H^s(\Omega)} \frac{(\tilde{\sigma}, E\varphi)_S}{\|E\varphi\|_{s,S}} \leq C\|\tilde{\sigma}\|_{-s}.$$

This completes the proof of Lemma 3.2.

Bibliographical Notes

Most of the work on multigrid methods for pseudodifferential operators has been concerned either with Fredholm operators of the second kind or hypersingular operators of the first kind. The second kind of operators are of the form of the identity plus a compact operator. At least in principal, such operators are already well conditioned so that it is not clear that much is to be gained by using a multigrid approach.

The hypersingular operators of the first kind are of positive order so that they fit into a framework similar to that of second order elliptic differential boundary problems. Work on this subject is contained in [142]. In contrast, weakly singular operators of potential theory give rise to pseudodifferential operators of negative order and require special consideration.

There has been a lot of work on boundary element methods. The paper [131] contains many results concerning Fredholm first kind equations arising in potential theory. The multigrid method, including the formulation of the method and its analysis, is taken from [29].

9. Application to fourth order problems

In Sections 6, 7 and 8 the examples considered all involved nested spaces and inherited forms. In this section we describe a practical setting in which the nonnested space, noninherited form multigrid theory of Section 4 may be profitably employed. Consider, on a planar domain Ω, the biharmonic Dirichlet problem.

(9.1)
$$\Delta^2 u = f$$
$$u = \frac{\partial u}{\partial n} = 0 \text{ on } \partial\Omega,$$

where $\frac{\partial}{\partial n}$ is the outward normal derivative operator on the boundary $\partial\Omega$ of Ω. For convenience we have assumed that the boundary conditions are homogeneous, but our arguments may be extended to include general Dirichlet boundary conditions. The problem (9.1) arises in models for clamped elastic plates and incompressible Stokes flow.

An important weak formulation of (9.1) seeks $u \in H_0^2(\Omega)$ such that

(9.2)
$$\sum_{i,j} \int_\Omega \frac{\partial^2 u}{\partial x_i \partial x_j} \frac{\partial^2 v}{\partial x_i \partial x_j} \, dx = \int_\Omega fv \, dx, \text{ for all } v \in H_0^2(\Omega).$$

It can be shown that the left hand side of (9.2) induces a norm on $H_0^2(\Omega)$. Hence, it follows from the Riesz Representation Theorem (cf. Section 2) that (9.2) has a unique solution for each $f \in H^{-2}(\Omega)$. Throughout this section we shall denote the dual space of $H_0^s(\Omega)$, $s > 0$, by $H^{-s}(\Omega)$.

By selecting a finite–dimensional (piecewise polynomial) test space M_h one may obtain a finite element method for (9.2) in the usual way. The condition $M_h \subset H_0^2$ leads to the requirement that C^1 finite elements be used. Since such elements are rather awkward to use in practice, other finite element approaches have been developed. In this section we will consider the behavior of Algorithm III when it is applied to two of these alternative methods. The first of these was described by

L.S.D. Morley [127]. When Morley nonconforming discretizations are constructed with respect to a sequence of nested triangulations, inherited quadratic forms and nonnested spaces are generated. Alternatively, when certain "mixed methods" (for example, the mixed method of Ciarlet and Raviart) are employed, nested spaces, but noninherited quadratic forms result.

Recall from Section 4 that if A.10 is satisfied and if smoothers satisfying A.4 are used, then it is possible to prove that a variable V–cycle preconditioner, defined by Algorithm III, leads to problems with uniformly bounded condition numbers (in the nonnested, noninherited setting) even if A.11 fails. Convergence of the W–cycle is also proved without A.11, provided that "m is sufficiently large". It is interesting to note that calculations performed by Hanisch showed that in practice it is often necessary to take $m > 1$ when using W–cycles to solve Morley or Ciarlet–Raviart equations. The condition A.11 fails in each case and W–cycle iterations for these methods with small m are observed to diverge. On the other hand, variable V–cycle iterations with $m = 1$ consistently provide excellent preconditioners for both problems.

It will be possible to verify Condition A.10 for the Morley and Ciarlet–Raviart methods so that Theorems 4.6 and 4.3 are applicable. For this it is convenient (but not necessary) to assume that Ω is a convex polygon. In this case, the solution u of (9.1) is known to belong to $H^3(\Omega) \cap H^2_0(\Omega)$ and to satisfy the a priori estimate

$$(9.3) \qquad\qquad \|u\|_3 \leq C\|f\|_{-1}$$

with C independent of $f \in H^{-1}(\Omega)$. Full regularity, $u \in H^4(\Omega)$, which can be proved if Ω has a smooth boundary, is not generally obtained on polygonal domains. The regularity of solutions for (9.1) on polygonal domains, including re–entrant corners and cracks, is studied in [69].

<u>Morley nonconforming finite element method</u>

Let $\mathcal{T}_h$ denote a triangulation, with mesh diameter h, of a convex polygon $\Omega \subset R^2$. The finite element spaces of Morley are defined so that $u \in M_h$ if and only if:

a) for each triangle $\tau \in T_h$, $u|_\tau$ is a quadratic polynomial,

b) u is continuous at triangle vertices and vanishes at boundary vertices, and

c) the normal derivative $\frac{\partial u}{\partial n}$ is continuous at the midpoints of each $\tau \in T_h$ and vanishes at midpoints along $\partial \Omega$.

First, it is clear that $M_h \not\subset C^0(\Omega)$. Consequently, if $T_{h/2}$ is defined from T_h by the usual halving strategy, then $M_h \not\subset M_{h/2}$. Morley spaces defined on a sequence of nested triangulations are themselves nonnested.

Consider next the bilinear forms on $M_{h_k} \times M_{h_k}$ defined by

$$A_k(u, v) = \sum_{\tau \in T_{h_k}} \sum_{i,j} \int_\tau \frac{\partial^2 u}{\partial x_i \partial x_j} \frac{\partial^2 v}{\partial x_i \partial x_j} \, dx.$$

The Morley method approximates (9.2) with the problem:

Find $u_k \in M_{h_k}$ such that for $f \in L^2(\Omega)$

(9.4) $$A_k(u_k, v) = (f, v), \text{ for all } v \in M_{h_k}.$$

It can be shown that $A_k(\cdot, \cdot)$ is an inner product for the space M_{h_k} so that (9.4) has a unique solution. Denote the induced norm

$$\|u\|_{2,h_k} \equiv A_k(u, u)^{1/2}.$$

We further observe that, for the nested meshes, $A_k(v, v) = A_J(v, v)$ for all v in the coarser space M_{h_k}.

It is possible to prove optimal approximation properties for the Morley spaces M_{h_k}. In addition, the following error estimate is known:

$$\|u - u_k\|_{2,h_k} \le C h_k [\|u\|_3 + h_k \|f\|_0]$$

where u and u_k denote the solutions of (9.2) and (9.4) respectively. The requirement $f \in L^2(\Omega)$ is necessary for (f, v) in (9.4) to make sense since the space M_{h_k} is discontinuous. Let v^I denote the piecewise T_{h_k}-linear interpolant of $v \in M_{h_k}$.

110

Then the Morley method may be extended to allow f in H^{-1}, (since $v^I \in H_0^1$), as follows:

Find $\bar{u}_k \in M_{h_k}$ such that

(9.5) $$A_k(\bar{u}_k, v) = (f, v^I), \quad \text{for all } v \in M_{h_k}.$$

The (unique) solution $\bar{u}_k$ of (9.5) satisfies the error estimate

$$\|u - \bar{u}_k\|_{2,h_k} \leq C h_k \|u\|_3.$$

Multigrid for the Morley method

In the usual way we construct a nested sequence of triangulations $T_k \equiv T_{h_k}$, $k = 1, \ldots, J$, of Ω. To obtain the Morley approximation u_J for the finest of these meshes, we consider the sequence of Morley discretizations associated with $\{T_{h_k}\}$ and apply the multigrid idea. The resulting quadratic forms $A_k(\cdot, \cdot)$ were observed to be "inherited", but the spaces $M_k \equiv M_{h_k}$ are nonnested.

The operators $A_k : M_k \to M_k$ defined from the Morley quadratic forms by

$$(A_k u, v) = A_k(u, v), \quad \text{for all } v \in M_k$$

are exceptionally ill–conditioned. It can be shown that $K(A_k) = O(h_k^{-4})$ as $h_k \to 0$. The bound for the maximum eigenvalue

(9.6) $$\lambda_k \leq C h_k^{-4}$$

is obtained using an inverse property for piecewise quadratic polynomials. That C is independent of k is a consequence of the quasi–uniformity of the mesh family $\{T_{h_k}\}$.

Assume that the smoothers of Section 5 are used. It remains to define an injection operator $I_k : M_{k-1} \to M_k$ for which A.10 can be proved. The following choice was analyzed in [56].

For $v \in M_{k-1}$, $I_k v \in M_k$ is defined so that

a) if p is a vertex of T_{k-1}, $(I_k v)(p) = v(p)$,

b) for other vertices p of T_k, v may have a jump at p and $I_k v$ takes the average value of v at p,

c) if m is a midpoint of an edge of T_k which is in the interior of $\tau \in T_{k-1}$, $\frac{\partial (I_k v)}{\partial n}(m) = \frac{\partial v}{\partial n}(m)$, and

d) for other edge midpoints m associated with T_k, $\frac{\partial v}{\partial n}$ may have a jump and $\frac{\partial (I_k v)}{\partial n}$ takes the average value of $\frac{\partial v}{\partial n}$ at m.

We may now consider Algorithm III with $p = 1$ and A.12 satisfied. To apply Theorem 4.6 we need to show that A.10 is satisfied for this choice of I_k. In [56] the following properties of I_k are proved

$$(9.7) \qquad \|I_k v\|_{2,h_k} \le C\|v\|_{2,h_{k-1}}, \text{ for all } v \in M_{k-1},$$

$$(9.8) \qquad \|u_k - I_k u_{k-1}\|_{2,h_k} \le C h_k[\|u\|_3 + h_k\|f\|_0]$$

and

$$(9.9) \qquad \|v^I - (I_k v)^I\|_1 \le C h_k \|v\|_{2,h_k}, \text{ for all } v \in M_{k-1},$$

where u_k and u_{k-1} are Morley approximations to the solution u of (9.2), and C is a constant independent of k. For the last of these bounds to make sense, v^I refers to the T_{k-1}–linear interpolant of v and $(I_k v)^I$ is the T_k–linear interpolant of $I_k v$.

Verification of A.10: Using the estimates (9.7), (9.8), and (9.9) it is possible to prove A.10 with $\alpha = \frac{1}{4}$ for the Morley method on a convex polygonal domain Ω. First, since

$$|A_k((I - I_k P_{k-1})u, u)| \le \|(I - I_k P_{k-1})u\|_{2,h_k}\|u\|_{2,h_k}$$

with $\|u\|_{2,h_k}^2 = A_k(u, u)$, it suffices to show that

$$\|(I - I_k P_{k-1})u\|_{2,h_k} \le C\lambda_k^{-1/4}\|A_k u\|_0^{1/2} A_k(u, u)^{1/4},$$

for all $u \in M_k$. In order to do this it will be useful to introduce, for any $u \in M_k$, the element $G_u \in M_k$ satisfying

$$A_k(u, v) = (G_u, v), \text{ for all } v \in M_k.$$

(Hence $G_u = A_k u$.) Denote by w the solution of the continuous problem (9.2) when $f = G_u$. (Since $H_0^1(\Omega)$ is continuously embedded in $L^2(\Omega)$, $(G_u, \cdot)$ defines an element of $H^{-1}(\Omega)$.) Further denote by w_k and w_{k-1} the Morley approximations to w on meshes T_k and T_{k-1}. By definition $u = w_k$. We also assume that the triangulations T_k are obtained from a coarse triangulation T_1 of the convex polygon Ω by the "halving strategy". Consequently, $h_{k-1} = 2h_k$ and the spaces M_k satisfy familiar inverse properties.

LEMMA 9.1. *In the setting of the previous paragraph*

$$\|(I - I_k P_{k-1})u\|_{2,h_k} \le C h_k \|G_u\|_{-1}$$

with constant C independent of k.

Proof: Since $u = w_k$, and using (9.7)

(9.10)
$$\|u - I_k P_{k-1} u\|_{2,h_k} = \|w_k - I_k w_{k-1} + I_k(w_{k-1} - P_{k-1}u)\|_{2,h_k}$$
$$\le \|w_k - I_k w_{k-1}\|_{2,h_k} + C\|w_{k-1} - P_{k-1}u\|_{2,h_{k-1}}.$$

According to (9.8) and (9.3) with $f = G_u \in M_k$, and using the inverse property

(9.11)
$$\|v\|_0 \le C h_k^{-1} \|v\|_{-1}, \text{ for all } v \in M_k$$

the first term on the right in (9.10) is appropriately bounded.

To estimate the remaining term in (9.10), observe that

$$A_{k-1}(w_{k-1}, v) = (G_u, v), \text{ for all } v \in M_{k-1}$$

and

$$A_{k-1}(P_{k-1}u, v) = A_k(u, I_k v) = (G_u, I_k v), \text{ for all } v \in M_{k-1}.$$

113

Hence

$$(9.12) \quad \|w_{k-1} - P_{k-1}u\|_{2,h_{k-1}} = \sup_{v \in M_{k-1}} \frac{A_{k-1}(w_{k-1} - P_{k-1}u, v)}{\|v\|_{2,h_{k-1}}}$$

$$= \sup_{v \in M_{k-1}} \frac{(G_u, v - I_k v)}{\|v\|_{2,h_{k-1}}}.$$

Adding and subtracting the T_{k-1}-linear interpolant v^I and the T_k-linear interpolant $(I_k v)^I$

$$(G_u, v - I_k v) = (G_u, v - v^I + v^I - (I_k v)^I + (I_k v)^I - I_k v)$$

$$\leq \|G_u\|_0 [\|v - v^I\|_0 + \|(I_k v)^I - I_k v\|_0] + \|G_u\|_{-1} \|v^I - (I_k v)^I\|_1.$$

Using the approximation properties of the piecewise linear interpolant and (9.7)

$$\|v - v^I\|_0 \leq C h_{k-1}^2 \|v\|_{2,h_{k-1}}$$

$$\|(I_k v)^I - I_k v\|_0 \leq C h_k^2 \|I_k v\|_{2,h_k} \leq C h_k^2 \|v\|_{2,h_{k-1}}.$$

Combining this with (9.9), (9.11), and since $h_{k-1} = 2h_k$

$$(G_u, v - I_k v) \leq C h_k \|G_u\|_{-1} \|v\|_{2,h_{k-1}},$$

which, with (9.12), completes the proof.

Since $h_k \leq C \lambda_k^{-1/4}$ follows from (9.6), it remains to prove that

$$(9.13) \quad \|G_u\|_{-1} \leq C \|A_k u\|_0^{1/2} A_k(u, u)^{1/4}.$$

To prove this, we introduce a scale of norms on M_k

$$\|u\|_{s,h_k} = \|A_k^{s/4} u\|_0.$$

Note that this agrees with the previous definition when $s = 2$. Using standard arguments it is possible to prove that

$$\|Q_k u\|_{2,h_k} \leq C \|u\|_2, \quad \text{for all } u \in H_0^2(\Omega)$$

114

where Q_k denotes the L^2-projection onto M_k. (To see this subtract u^I, use an inverse property and then apply standard error estimates.) Clearly,

$$\|Q_k u\|_{0,h_k} = \|Q_k u\|_0 \le \|u\|_0, \text{ for all } u \in L^2(\Omega).$$

By the interpolation, Theorem B.4, we have

$$\|Q_k u\|_{1,h_k} \le C\|u\|_1, \text{ for all } u \in H_0^1(\Omega).$$

To complete the proof of A.10, observe that

$$\|G_u\|_{-1} = \sup_{v \in H_0^1(\Omega)} \frac{(G_u, v)}{\|v\|_1} \le C \sup_{v \in H_0^1(\Omega)} \frac{(G_u, v)}{\|Q_k v\|_{1,h_k}}.$$

Since $G_u \in M_k$

$$(G_u, v) = (G_u, Q_k v) \le \|G_u\|_{-1,h_k} \|Q_k v\|_{1,h_k}$$

by the Cauchy–Schwarz inequality. Consequently, $\|G_u\|_{-1} \le C\|G_u\|_{-1,h_k}$. As a further consequence of the Cauchy–Schwarz inequality

$$\|G_u\|_{-1,h_k} \le \|G_u\|_{0,h_k}^{1/2} \|G_u\|_{-2,h_k}^{1/2}$$

and (9.13) is an immediate consequence of the identification, $G_u = A_k u$,

$$\|G_u\|_{0,h_k} = \|A_k u\|_{0,h_k} = \|A_k u\|_0$$

and

$$\|G_u\|_{-2,h_k} = \|A_k^{-1/2} G_u\|_0 = \|A_k^{1/2} u\|_0 = A_k(u, u)^{1/2}.$$

This completes the verification of A.10.

A second weak formulation of (9.1). In order to formulate the mixed method of Ciarlet and Raviart we need an additional weak formulation of (9.1). This may be obtained by introducing an auxiliary variable σ satisfying

$$\sigma = -\Delta u,$$

and

$$-\Delta\sigma = f.$$

Multiplying by test functions and integrating by parts, we are led to the following mixed–variable weak formulation.

Find $\{u,\sigma\} \in H_0^1(\Omega) \times H^1(\Omega)$ such that

$$(\sigma, v) - D(v, u) = 0, \text{ for all } v \in H^1(\Omega)$$

and

$$-D(\sigma, w) = -(f, w), \text{ for all } w \in H_0^1(\Omega)$$

where $D(\cdot, \cdot)$ denotes the Dirichlet form

$$D(v, u) = \int_\Omega \nabla v \cdot \nabla u \, dx.$$

It can be shown that when Ω is a convex polygon, this problem has a unique solution $\{u, \sigma\}$, and that u solves (9.1) and (9.2), and that $\sigma = -\Delta u$. (If Ω is such that H^3–regularity is not obtained, for example, when Ω has a re–entrant corner, it is necessary to change the test spaces in the above formulation as is described in [99].)

The mixed method of Ciarlet and Raviart

A finite element method may be obtained by taking finite dimensional test spaces $M_h \subset H_0^1(\Omega)$ and $\tilde{M}_h \subset H^1(\Omega)$ in the above formulation. To obtain uniqueness and good error estimates, the spaces M_h and $\tilde{M}_h$ must be "complementary". More precisely, M_h and $\tilde{M}_h$ must satisfy the so–called "inf–sup" conditions(cf. [7] or [57]). For example, one may choose $\tilde{M}_h$ to be the space of continuous piecewise polynomials of degree less than or equal to m (defined with respect to a triangulation T_h of the convex polygon Ω) with $M_h = \tilde{M}_h \cap H_0^1(\Omega)$. The resulting finite element methods were first analyzed by P. Ciarlet and P.–A. Raviart. If $\{u_h, \sigma_h\}$ denotes the mixed finite element solution, and if $m \geq 2$, the following error estimates are known

$$\|\sigma - \sigma_h\|_0 \leq Ch\|\sigma\|_1,$$

116

and

$$\|u - u_h\|_1 \le Ch^2 \|u\|_3.$$

These error estimates are used in a proof of A.10 in [99]. Hence we assume that $m \ge 2$ which means that continuous piecewise quadratics (at least) are used.

We next consider the resulting system of linear algebraic equations. Given bases $\{\varphi^i\}$ and $\{\tilde{\varphi}^i\}$ for M_h and $\tilde{M}_h$, respectively, the Ciarlet–Raviart method reduces to a matrix equation for the coefficient vectors U_h of u_h and Σ_h of σ_h. If we denote $[\mathcal{I}_h]_{ij} = (\tilde{\varphi}^i, \tilde{\varphi}^j)$, $[D_h]_{ij} = -D(\tilde{\varphi}^j, \varphi^i)$, and $[F_h]_i = (f, \varphi^i)$, then

$$(9.14) \qquad \mathcal{N}_h \begin{pmatrix} \Sigma_h \\ U_h \end{pmatrix} \equiv \begin{pmatrix} \mathcal{I}_h & D_h^T \\ D_h & 0 \end{pmatrix} \begin{pmatrix} \Sigma_h \\ U_h \end{pmatrix} = \begin{pmatrix} 0 \\ -F_h \end{pmatrix}.$$

Since the matrix $\mathcal{N}_h$ is symmetric but indefinite, it is convenient to solve the Schur complement equation (obtained by eliminating Σ_h in (9.14))

$$(9.15) \qquad S_h U_h \equiv D_h \mathcal{I}_h^{-1} D_h^T U_h = F_h.$$

The vector Σ_h can be obtained from U_h by back–solving. The Schur complement S_h is SPD but, like $\mathcal{N}_h$, it is ill–conditioned with condition number $K(S_h) = O(h^{-4})$ as $h \to 0$.

Note that in order to compute the action of the Schur complement S_h it is necessary to "invert" the L^2 Gram matrix $\mathcal{I}_h$. For the usual choices of the basis $\{\tilde{\varphi}^i\}$, $\mathcal{I}_h$ is well–conditioned and the action of its inverse can be computed with a rapidly converging iterative process. Alternatively, it is possible to replace the L^2 inner product in the mixed method equations with certain approximate L^2 inner products $(\cdot, \cdot)_h$. The solution $\{\bar{u}_h, \bar{\sigma}_h\}$ of this perturbed mixed method exhibits similar approximation properties. Furthermore, a new basis $\{\bar{\varphi}^i\}$ can be constructed for which the matrix with entries $(\bar{\varphi}^i, \bar{\varphi}^j)_h$ is diagonal and $D(\bar{\varphi}^j, \varphi^i)$ remains sparse. This approach is described in [99].

Multigrid for the Ciarlet–Raviart mixed method

117

Consider a sequence of Ciarlet–Raviart discretizations, one for each mesh in the usual family of nested triangulations $\{T_{h_k}\}$, $k = 1, \dots, J$ of Ω. Observe that the resulting spaces $M_k \equiv M_{h_k}$ are nested. Consequently, in defining a multigrid algorithm to solve for $u_J \equiv u_{h_J}$, we may set $I_k : M_{k-1} \to M_k$ equal to the natural injection operator.

Let $< \cdot, \cdot >_k$ denote the Euclidean inner product on $R^n \times R^n$ with $n = N_k$, the dimension of M_k. Corresponding to each space M_k one has the natural quadratic form

$$A_k(u, v) = < S_{h_k} U, V >_k$$

where U, V denote the coefficient vectors of $u, v \in M_k$ (with respect to the basis $\{\varphi^i\}$). Note , with $D_k \equiv D_{h_k}$ and $I_k \equiv I_{h_k}$, that

(9.16)
$$A_k(u, u) = < D_k I_k^{-1} D_k^T U, U >_k$$

$$= < I_k I_k^{-1} D_k^T U, I_k^{-1} D_k^T U >_k$$

$$= \sup_{V \in R^n} \frac{< I_k I_k^{-1} D_k^T U, V >_k^2}{< I_k V, V >_k}$$

$$= \sup_{v \in \tilde{M}_k} \frac{D^2(v, u)}{\|v\|_0^2},$$

so that $A_k(\cdot, \cdot)$ is independent of the basis selected for M_k. On the other hand, the forms $A_k(\cdot, \cdot)$ are noninherited as the supremum in (9.16) is k–dependent. In fact, A.11 seems to fail in this setting.

In [99] Condition A.10 is proved for the Ciarlet–Raviart method. (It is quite tedious and will not be included here.) Consequently, using the smoothers of Section 5, Theorems 4.6 and 4.3 may be applied to show that B_J defined by Algorithm III, with $p = 1$ and $m(k)$ chosen to satisfy A.12, is a good preconditioner for $A..$

Bibliographical Notes

For an introduction to finite element methods for the biharmonic Dirichlet problem, see [65]. An extensive catalog of such methods is provided in [102]. Regularity results for the biharmonic problem are given in [69].   .

118

The Morley method was introduced in [127]. Error estimates and other analytical details can be found in [2], [109], [143] and [144]. Proofs of the estimates (9.7), (9.8) and (9.9) are given by Brenner in [56]. This paper also contains an analysis of the multigrid W–cycle in the Morley setting. Alternative multigrid approaches for solving the Morley equations can be found in [141] and [99].

Various analyses of the Ciarlet–Raviart method exist, see for example [66], [59], [83], [9] and [99]. The multigrid approach considered here is examined in detail in [99]. An alternative multigrid approach which solves (9.14) directly is analyzed by Peisker [140] and is based on the work of Verfürth [156].

10. Implementation, work estimates and full multigrid

Implementation issues

In the previous sections the approximate problems have been stated as operator equations on M. As in the application to second order problems (Sections 6 and 7) and to pseudodifferential operators (Section 8), the problems are presented as: Find $U \in M$ such that

$$A(U, \chi) = F(\chi)$$

or, find $u \in M$ such that

$$V(u, \chi) = F(\chi), \text{ for all } \chi \in M.$$

Most of the following discussion applies to both cases above and we shall use as generic notation that of the first example except when otherwise noted. However, before proceeding, we briefly examine the way finite element equations are described in computer codes. Let $\{\varphi_k^i\}$, $i = 1, \ldots, N_k$, denote a computational basis for the finite element space M_k. Typically, one represents unknown functions in M_k by a vector of coefficients with respect to the computational basis. On the other hand, the operator A_k defined by (2.10) is seldom directly represented on the computer. Instead, it is more convenient to compute the stiffness matrix

$$(\tilde{A}_k)_{ij} = A(\varphi_k^i, \varphi_k^j).$$

Problem (6.3) is then replaced by the matrix equation

(10.1) $$\tilde{A}_k \tilde{U} = \tilde{F}.$$

Here $\tilde{U}$ denotes the vector of coefficients of U in the computational basis and $\tilde{F}_i = F(\varphi_j^i)$ is a known data vector.

Let $g \in M$ be the unique function satisfying

$$(g, \theta) = F(\theta), \text{ for all } \theta \in M.$$

Then, the vector of coefficients $\tilde{g}$ for g in the computational basis is given by

$$\tilde{g} = \tilde{G}^{-1}\tilde{F}$$

where $\tilde{G}$ denotes the Gram matrix $\tilde{G}_{ij} = (\varphi_J^i, \varphi_J^j)$. In terms of the operator A_J, (6.3) can be written as

(10.2) $$A_J U = g.$$

To apply a preconditioned iterative method to solve (10.2), it appears necessary to compute the coefficients of g, i.e., to apply the action of the inverse of the Gram matrix $\tilde{G}$. This is not the case provided that the action of the preconditioner B_J applied to an arbitrary function v can be computed directly from the inner product data vector $\{(v, \varphi_J^i)\}$.

Thus, the problem of implementing the multigrid algorithm reduces to computing the coefficients of $B_J v$ (with respect to the computational basis for M) for an arbitrary vector v, given the corresponding inner product data $\{(v, \varphi_J^i)\}$. The action of the operator B_1 is computed from the inner product data by direct inversion of the matrix $\tilde{A}_1$. The computational algorithm is defined by mathematical induction. We will describe how to compute the action of B_k on v, given the inner product data $\{(v, \varphi_k^i)\}$, assuming that the action of B_{k-1} can be computed with similar data given.

For Steps 1 and 3 of the V–cycle algorithm, R_k and R_k^t must be such that the coefficients in the computational basis for M_k corresponding to both $R_k w$ and $R_k^t w$ for arbitrary vectors w must be computable given the inner product data $\{(w, \varphi_k^i)\}$. Note that A_k applied to x_2 in Step 3 is replaced by the multiplication of the coefficients of x_2 in the implementation by $\tilde{A}_k$ and results in the inner product data $\{(A_k x_2, \varphi_k^i)\}$.

Two additional ingredients are required for Step 2. To compute the coefficients of q with respect to the $k-1$'st basis, we must apply B_{k-1} and hence compute the inner product data corresponding to the function $Q_{k-1}(A_k x_1 - g)$. By the above considerations, the inner product data $\{(A_k x_1 - g, \varphi_k^i)\}$ are known. But, since $M_{k-1} \subset M_k$, the basis functions of M_{k-1} are expressible in terms of those of M_k. Hence

$$(Q_{k-1}(A_k x_1 - g), \varphi_{k-1}^i) = (A_k x_1 - g, \varphi_{k-1}^i)$$
$$= \sum_{j=1}^{N_k} \alpha_{ij}^k (A_k x_1 - g, \varphi_k^j).$$

Here $\varphi_{k-1}^i = \sum_{j=1}^{N_k} \alpha_{ij}^k \varphi_k^j$. The process of computing the inner product data with respect to the basis for M_{k-1} from the M_k inner product data corresponds to restriction in classical multigrid expositions. Note that, in practice, it often happens that α_{ij}^k is nonzero for only a few values of i and j.

Finally, we need to compute the coefficients of x_2, given those of x_1 and q corresponding to the computational basis for M_k and M_{k-1} respectively. The coefficients for q with respect to the computational basis for M_k can be computed from those for M_{k-1} by applying the transpose of the matrix $(\tilde{C}_{k-1})_{ij} = \alpha_{ij}^k$ above. This step corresponds to interpolation in classical multigrid expositions.

Instead of noting that a preconditioned iteration involving the operators B_J and A_J can be implemented, we can alternatively develop a matrix $\tilde{B}_J$ (coming from the multigrid algorithm) which preconditions the matrix $\tilde{A}_J$. The conditions on R_k and R_k^t are exactly the same as required above. That is, they must be such that the coefficients in the computational basis for M_k corresponding to both $R_k w$ and $R_k^t w$ for arbitrary vectors w are computable, given the inner product data $\{(w, \varphi_k^i)\}$. The corresponding linear processes which take vectors $\{(w, \varphi_k^i)\}$ into the coefficients for $R_k w$ and $R_k^t w$ will be denoted by the matrices $\tilde{R}_k$ and $\tilde{R}_k^t$ respectively. The matrix $\tilde{B}_J$ is defined in the following algorithm.

Algorithm M (Matrix form of the V-cycle algorithm)
Set $\tilde{B}_1 = \tilde{A}_1^{-1}$. Assume that $\tilde{B}_{k-1}$ has been defined and define $\tilde{B}_k \tilde{G}$ for $\tilde{G} \in R^{N_k}$ as follows:

(1) Set

$$\tilde{x}_1 = \tilde{R}_k^t \tilde{G}.$$

(2) Define $\tilde{x}_2 = \tilde{x}_1 - \tilde{C}_{k-1}^t \tilde{q}$ where

$$\tilde{q} = \tilde{B}_{k-1} \tilde{C}_{k-1} (\tilde{A}_k \tilde{x}_1 - \tilde{G}).$$

(3) Finally, set

$$\tilde{B}_k \tilde{G} = \tilde{x}_2 - \tilde{R}_k (\tilde{A}_k \tilde{x}_2 - \tilde{G}).$$

PROPOSITION 10.1. *Let $v \in M_k$ and $\tilde{v}$ denote the vector of coefficients for the expansion of the function v in the basis $\{\varphi_k^i\}$. Then, $\tilde{B}_k \tilde{A}_k \tilde{v}$ is the vector of coefficients for the expansion of the function $B_k A_k v$.*

PROOF: The proof is by induction. The result is obvious for $k = 1$. Let k be greater than one and let v and $\tilde{v}$ be as above. We consider applying the operator algorithm to $A_k v$ and the matrix algorithm to $\tilde{A}_k \tilde{v}$. In the first step of the operator algorithm, we compute the function $x_1 = R_k^t g$ for $g = A_k v$. In contrast, $\tilde{G}_i = (\tilde{A}_k \tilde{v})_i = (A_k v, \varphi_k^i)$ in the matrix algorithm. Thus, by the definition of $\tilde{R}_k^t$, $\tilde{x}_1 = \tilde{R}_k^t \tilde{G}$ is the vector of coefficients for the expansion of the function x_1.

To compare the results of the second steps of the algorithms, we note that the previous conclusion immediately implies that $\tilde{A}_k \tilde{x}_1 - \tilde{G} = \tilde{A}_k (\tilde{x}_1 - \tilde{v})$ is equal to the vector $\{(A_k(x_1 - v), \varphi_k^i)\}$. Consequently, $\tilde{C}_{k-1}(\tilde{A}_k \tilde{x}_1 - \tilde{G})$ is the vector $\{(A_{k-1}(x_1 - v), \varphi_{k-1}^i)\}$. However,

$$(10.3) \qquad \begin{aligned} (A_k(x_1 - v), \varphi_{k-1}^i) &= (Q_{k-1} A_k(x_1 - v), \varphi_{k-1}^i) \\ &= (A_{k-1} P_{k-1}(x_1 - v), \varphi_{k-1}^i). \end{aligned}$$

Note that the right hand side of equation (10.3) is $\tilde{A}_{k-1}$ applied to the coefficients of $P_{k-1}(x_1 - v)$ expanded in the basis for M_{k-1}. Thus, by the induction hypothesis, $\tilde{q}$ in the matrix algorithm is the vector of coefficients (with respect to the basis for M_{k-1}) of the expansion of

$$B_{k-1} A_{k-1} P_{k-1}(x_1 - v) = q.$$

Consequently, $\tilde{x}_2$ defined by the matrix algorithm gives the coefficients of x_2 defined by the operator algorithm.

The proof that the final step of the matrix algorithm results in the coefficients of the function developed in the final step of the operator algorithm is similar. This completes the proof of the proposition.

REMARK 10.1: The proposition immediately implies that for all $v \in M_J$,

$$< \tilde{A}_J(I - \tilde{B}_J \tilde{A}_J \tilde{v}), \tilde{v} >= A((I - B_J A_J)v, v)$$

and, by definition,

$$< \tilde{A}_J \tilde{v}, \tilde{v} >= A(v, v).$$

Here $< \cdot, \cdot >$ is the Euclidean inner product on R^{N_k} with $N_k = dim(M_k)$. Thus, contraction estimates for $I - B_J A_J$ and estimates on the condition number $K(B_J A_J)$ lead to the same results for their matrix counterparts.

The additive smoother defined by (5.1) of Section 5 is such that the coefficients of $R_k g$ are given by a sparse block matrix applied to the vector of inner product data $\{(g, \varphi_k^i)\}$. For simplicity, let the spaces be given by $M_k^i = span\{\varphi_k^i\}$, i.e., one dimensional spaces. Then

$$R_k g = \sum_{i=1}^{N_k} A_{k,i}^{-1} Q_k^i g,$$

and $A_{k,i}^{-1} Q_k^i g = c_i \varphi_k^i$. Thus the problem reduces to computing c_i. But

$$c_i A(\varphi_k^i, \varphi_k^i) = A(A_{k,i}^{-1} Q_k^i g, \varphi_k^i) = (g, \varphi_k^i).$$

Hence

$$c_i = (A(\varphi_k^i, \varphi_k^i))^{-1}(g, \varphi_k^i).$$

The argument showing that the coefficients of the result of the multiplicative smoother can be computed from the corresponding inner product data is similar and can be found in [34].

124

In a similar way we can have more general subspace decompositions

$$M_k = \sum_{i=1}^{\ell_k} M_k^i.$$

Instead of computing the coefficients c_i we need, for each M_k^i, to invert the matrix $\{A(\varphi_k^\ell, \varphi_k^m)\}$ where we have assumed that M_k^i has a basis consisting of basis functions coming from the basis $\{\varphi_k^i\}$, and ℓ and m range over those indices. Thus we could have overlapping spaces, or a direct sum if the spaces are essentially disjoint. This includes the additive and multiplicative "block" methods or, what is the same, the block Jacobi and Gauss–Seidel methods.

We finally turn to the case of Section 8. Here $V(\cdot, \cdot)$ and $< \cdot, \cdot >_{-1}$ replace $A(\cdot, \cdot)$ and $(\cdot, \cdot)$ in the earlier discussion. We need only show how to compute the coefficients of $R_k g$ in the computational basis from the inner product data $\{< g, \varphi_k^i >_{-1}\}$.

By the definition of R_k,

(10.4) $$[R_k g, \varphi_k^i]_k = < g, \varphi_k^i >_{-1} .$$

Combining (8.7) and (10.4), we see that

$$(R_k g, \varphi_k^i)_k = [R_k g, A_k \varphi_k^i]_k = < g, A_k \varphi_k^i >_{-1} \qquad \text{for } i = 1, \ldots, N_k.$$

Using the fact that

(10.5) $$(\varphi_k^i, \varphi_k^j)_k = \delta_{ij},$$

where δ_{ij} is the Kronecker delta, it follows that

$$R_k g = \sum_{i=1}^{N_k} < g, A_k \varphi_k^i >_{-1} \varphi_k^i$$

$$= \sum_{i,j=1}^{N_k} < g, \varphi_k^j >_{-1} A(\varphi_k^i, \varphi_k^j)\varphi_k^i.$$

125

Thus,

$$\sum_{j=1}^{N_k} < g, \varphi_k^j >_{-1} A(\varphi_k^i, \varphi_k^j)$$

gives the i'th coefficient of $R_k g$ in terms of the inner product data.

The smoothers of Section 5 do not appear to fit into the case treated in Section 8.

Work estimates

We continue this section with some work estimates. For this purpose let $N_k = \dim(M_k)$ and suppose that for some $a > 1$

(10.6) $$N_{k+1} \geq aN_k.$$

We also assume that for $\beta_1 \geq 1$

(10.7) $$m(k-1) \leq \beta_1 m(k).$$

Let p be the parameter of the general multigrid algorithm of Section 4. Let W_k be the work to evaluate B_k. We assume that the work to evaluate A_k and R_k is proportional to N_k. This is true for the example in Section 6, but may not hold for the example of Section 8. We suppose that for the coarse grid we have

(10.8) $$W_1 \leq C_1 N_1$$

for some constant C_1. Now we have, by examining the algorithm and ignoring the cost of intergrid transfers

(10.9) $$W_k \leq pW_{k-1} + C_2 m(k)N_k$$

for some constant C_2. We will always assume that $\frac{p}{a}\beta_1 < 1$. Without such a condition the work could increase exponentially.

By examining the recurrence (10.9) under the conditions (10.6), (10.7) and (10.8) we are led to the following:

(10.10) $$W_k \leq C_1 N_1 \left(\frac{N_k}{N_1}\right)^{\frac{\log p}{\log a}} + \left(1 - \frac{p}{a}\beta_1\right)^{-1} C_2 m(k)N_k.$$

126

That this is true is easily seen by induction. Clearly it is true for $k = 1$. Assuming it is true for $k - 1$, we have from (10.9), (10.6) and (10.7)

$$W_k \leq pC_1N_1\left(\frac{N_{k-1}}{N_1}\right)^{\frac{\log p}{\log a}} + p\left(1 - \frac{p}{a}\beta_1\right)^{-1}C_2m(k-1)N_{k-1} + C_2m(k)N_k$$

$$\leq C_1N_1\left(\frac{N_k}{N_1}\right)^{\frac{\log p}{\log a}}pa^{-\frac{\log p}{\log a}} + C_2\left[\frac{p}{a}\beta_1\left(1 - \frac{p}{a}\beta_1\right)^{-1} + 1\right]m(k)N_k,$$

which is (10.10).

We will focus our attention on the example of Section 6 where a halving strategy was used. Thus a is approximately 4. We will assume that $a = 4$. Now we need $p\beta_1 < 4$.

In the case of the V–cycle with $p = 1$, $m(k) = m$ and $\beta_1 = 1$, (10.10) is

$$W_J \leq C_1N_1 + \frac{4}{3}C_2mN_J.$$

Now C_2mN_J is essentially the cost of $2m$ evaluations of A_J and hence, neglecting C_1N_1, the total cost for one V–cycle with m smoothings is less than $3m$ evaluations of A_J.

For the next example we consider the variable V–cycle with $p = 1$, $\beta_1 = 2$, $a = 4$ and $m(J) = 1$. Then (10.10) yields

(10.11) $$W_J \leq C_1N_1 + 2C_2N_J.$$

Hence the cost is approximately that of 4 evaluations of A_J.

Finally we consider the W–cycle, i.e., we take $p = 2$, $\beta_1 = 1$ and $m = 1$. Then in this case we get an estimate with the same second term as in (10.11). The W–cycle does involve more intergrid transfer operations which we have ignored in this discussion.

Full multigrid

We shall consider here the example of Section 6. In such an example, error estimates for the approximate solution (6.3) of (6.2) are well known. On each level k we define the kth level approximate solution by

$$A(U_k, \chi) = (f, \chi), \quad \text{for all } \chi \in M_k.$$

Now the following error estimate is standard:

$$(10.12) \qquad \|u - U_k\|_1 \leq Ch_k^\alpha \|f\|_{-1+\alpha},$$

where $\alpha \in (0, 1]$ as in (6.4).

The so-called full multigrid algorithm starts with $U_1 \in M_1 \subset M_2$ as an approximation $\tilde{U}_2 = U_1$ of U_2. After n steps of some (efficient) algorithm, using U_1 as a starting guess, we produce a better approximation U_2^n satisfying

$$\|U_2 - U_2^n\|_1 \leq \delta^n \|U_2 - \tilde{U}_2\|_1.$$

The two level multigrid iteration is an example of such a process. Now for $k > 2$ set $\tilde{U}_k = U_{k-1}^n$ and find an approximate solution U_k^n to U_k satisfying

$$(10.13) \qquad \|U_k - U_k^n\|_1 \leq \delta^n \|U_k - U_{k-1}^n\|_1.$$

A k-level multigrid process could be used here, for example.

Now we may combine the estimates to show that, for some constant $\tilde{C}$,

$$(10.14) \qquad \|u - U_J^n\|_1 \leq \tilde{C}h^\alpha \|f\|_{-1+\alpha}.$$

To establish (10.14) we proceed as follows: By (10.13)

$$\|u - U_J^n\|_1 \leq \|u - U_J\|_1 + \|U_J - U_J^n\|_1$$
$$\leq \|u - U_J\|_1 + \delta^n \|U_J - U_{J-1}^n\|_1$$
$$\leq (1 + \delta^n)\|u - U_J\|_1 + \delta^n \|u - U_{J-1}^n\|_1.$$

Clearly, continuation of this argument implies that

$$\|u - U_J^n\|_1 \leq 2 \sum_{k=0}^{J-1} \delta^{nk} \|u - U_{J-k}\|_1.$$

Using (10.12), we have

$$\|u - U_J^n\|_1 \leq 2C \Big(\sum_{k=0}^{J-1} h_{J-k}^\alpha \delta^{nk} \Big) \|f\|_{-1+\alpha}.$$

Now $h_{J-k} = 2^k h_J = 2^k h$. Hence

$$\sum_{k=0}^{J-1} h^\alpha_{J-k} \delta^{nk} = h^\alpha \sum_{k=0}^{J-1} (2^\alpha \delta^n)^k.$$

Thus if n is chosen so that $2^\alpha \delta^n \le \eta < 1$ then

$$\sum_{k=0}^{J-1} (2^\alpha \delta^n)^k \le \frac{1}{1-\eta}.$$

Therefore we have that

$$\|u - U_J^n\|_1 \le 2C(1-\eta)^{-1} h^\alpha \|f\|_{-1+\alpha},$$

which is (10.14).

The importance of this is that if $\delta < 1$ is independent of h (i.e., independent of the number of levels) then n may be chosen to be a fixed constant independent of h. Now we see that the total cost of arriving at an approximation of u is the sum of the costs on all levels. Hence if $\overline{W}_k$ is the cost on level k of one iterative step (say using the multigrid V–cycle) then the total cost W is

$$W = n \sum_{k=1}^{J} \overline{W}_k.$$

Now suppose $\overline{W}_k \le \overline{C} N_k$. Then

$$W \le \overline{C} n \sum_{k=1}^{J} N_k \le \overline{C} n N_J \sum_{k=0}^{J} 4^{-k} \le (4/3) n \overline{C} N_J.$$

Hence the full multigrid process is said to have "optimal complexity", i.e., the number of operations required to obtain the desired approximation is proportional to the number of unknowns.

Bibliographical Notes

The work estimates are essentially the same as those given in [12]. In that paper they also considered the "full multigrid" process and proved that the W–cycle led to an algorithm of optimal complexity.

In some works, the multigrid algorithms are defined with a simple scaling of the identity operator as the smoother (cf. [12]). In such cases, discrete inner products, varying from level to level, are introduced in order to obtain an easily computable implementation of the algorithm. The implementation issues discussed here are discussed in [34] and [29] where it was shown that, with the smoother properly defined, the same inner product can be used on all levels to define the relevant operators, while still maintaining the ease of implementation.

Appendix A: The Conjugate Gradient Algorithm

In Section 1, the conjugate gradient algorithm (CG) for the approximation of the solution, $x \in M$, of

(a.1)
$$Ax = b,$$

where A is SPD, was briefly discussed. More precisely it is given as follows: Let x_0 be an initial guess. Set $r_0 = b - Ax_0$ and define

$$V_n = span\{r_0, Ar_0, A^2r_0, \ldots, A^{n-1}r_0\}.$$

(This is called the Krylov subspace of M relative to r_0 and A.) The nth CG iterate, x_n, is defined in terms of the Galerkin approximation $\hat{x}_n$ to $x - x_0$ in V_n; i.e., for $n = 1, 2, \ldots$

(a.2)
$$(A\hat{x}_n, \varphi) = (r_0, \varphi), \text{ for all } \varphi \in V_n.$$

Define $x_n = \hat{x}_n + x_0$. Then (a.2) is the same as

(a.3)
$$(Ax_n, \varphi) = (b, \varphi),$$

for all $\varphi \in V_n$ and $x_n - x_0 \in V_n$.

If B is another SPD operator, then (a.1) is equivalent to

(a.4)
$$BAx = Bb.$$

Now defining the new inner product on M by $[\cdot, \cdot] = (B^{-1}\cdot, \cdot)$, we see that $\mathcal{A} = BA$ is symmetric with respect to $[\cdot, \cdot]$, i.e., $[\mathcal{A}x, y] = [x, \mathcal{A}y]$. Clearly $\mathcal{A}$ is positive definite. The preconditioned conjugate gradient algorithm (PCG) is now the following: Let y_0 be an initial guess. Set $z_0 = Bb - \mathcal{A}y_0$ and define

$$\tilde{V}_n = span\{z_0, \mathcal{A}z_0, \mathcal{A}^2 z_0, \ldots, \mathcal{A}^{n-1}z_0\}.$$

(This is called the Krylov subspace of M relative to z_0 and $\mathcal{A}$.) The nth PCG iterate, y_n, is defined in terms of the Galerkin approximation $\hat{y}_n$ to $x - y_0$ in $\tilde{V}_n$; i.e., for $n = 1, 2, \ldots$

(a.5) $[\mathcal{A}\hat{y}_n, \varphi] = [z_0, \varphi]$, for all $\varphi \in \tilde{V}_n$.

Define $y_n = \hat{y}_n + y_0$. Then (a.5) is the same as

(a.6) $(Ay_n, \varphi) = (b, \varphi)$,

for all $\varphi \in \tilde{V}_n$ and $y_n - y_0 \in \tilde{V}_n$. Comparing (a.2) and (a.5), we see that PCG is formally parallel to CG if we substitute $\mathcal{A}$ for A and $[\cdot, \cdot]$ for $(\cdot, \cdot)$. Noting (a.6), another way of looking at PCG is that $y_n - y_0$ is just the usual Galerkin approximation to $x - y_0$ in the subspace $\tilde{V}_n$. If we choose $B = I$, then $\tilde{V}_n \equiv V_n$ so that CG is a special case of PCG.

From the standpoint of speed of convergence, as an iterative algorithm, the preconditioner B plays a major role. We will see this from the error estimate which follows.

Remark. By the Cayley–Hamilton theorem, A^{-1} is a polynomial in A of degree less than or equal to $N - 1$ where $N = \dim(M)$. Hence $x - x_0 = A^{-1}r_0 \in V_N$. Therefore (with exact computations) CG reaches the desired solution in no more than N steps. The conjugate gradient algorithm is much more important, however, when considered as an iterative method since, typically, N is much larger than the number of steps required to reach a desired accuracy. For example, in finite element computation, N may be of the order of 100,000 whereas, if a good preconditioner B is chosen, $n = 10$ or 20 may suffice.

Analysis of the error of PCG

We shall estimate the error in the norm $[\mathcal{A}\cdot, \cdot]^{1/2} = (A\cdot, \cdot)^{1/2}$. From (a.5) it follows that

(a.7) $[\mathcal{A}(\hat{y}_n - (x - y_0)), \varphi] = 0$, for all $\varphi \in \tilde{V}_n$.

132

Hence setting $e_n = \hat{y}_n - (x - y_0) = y_n - x$ it follows that

(a.8) $$[\mathcal{A}e_n, e_n] = [\mathcal{A}e_n, y - (x - y_0)],$$

for all $y \in \tilde{V}_n$. Hence using the Cauchy–Schwarz inequality in $[\mathcal{A}\cdot, \cdot]$, we have

(a.9) $$[\mathcal{A}e_n, e_n] \leq [\mathcal{A}(y - (x - y_0)), y - (x - y_0)]$$

for any $y \in \tilde{V}_n$. Now since $y \in \tilde{V}_n$, we may choose an arbitrary polynomial $\mathcal{P}_{n-1}$ of degree $n-1$ and take

$$y = \mathcal{P}_{n-1}(\mathcal{A})z_0 \equiv \mathcal{P}_{n-1}(\mathcal{A})\mathcal{A}(x - y_0).$$

With this choice (a.9) becomes

(a.10) $$[\mathcal{A}e_n, e_n] \leq [\mathcal{A}(I - \mathcal{P}_{n-1}(\mathcal{A})\mathcal{A})(x - y_0), (I - \mathcal{P}_{n-1}(\mathcal{A})\mathcal{A})(x - y_0)].$$

Set $I - \mathcal{P}_{n-1}(\mathcal{A})\mathcal{A} = \mathcal{Q}_n(\mathcal{A})$. Then

(a.11) $$[\mathcal{A}\mathcal{Q}_n(\mathcal{A})(x - y_0), \mathcal{Q}_n(\mathcal{A})(x - y_0)] \leq [[\mathcal{Q}_n(\mathcal{A})]]^2[\mathcal{A}(x - y_0), x - y_0],$$

where $[[\cdot]]$ is the operator norm induced by $[\cdot, \cdot]$. Hence (a.10) and (a.11) imply

(a.12) $$(\mathcal{A}e_n, e_n) \leq [[\mathcal{Q}_n(\mathcal{A})]]^2(\mathcal{A}(x - y_0), x - y_0)$$

for any polynomial $\mathcal{Q}_n$ of degree n with $\mathcal{Q}_n(0) = 1$. Since $\mathcal{A}$ is SPD, its spectrum is real and positive. Let $\underline{\lambda}$ and $\overline{\lambda}$ be the extreme eigenvalues of $\mathcal{A}$ and set $\mathcal{M} = I - \frac{2}{\overline{\lambda} + \underline{\lambda}}\mathcal{A}$. For each polynomial $\tilde{\mathcal{Q}}_n(t)$ of degree n with $\tilde{\mathcal{Q}}_n(1) = 1$ set $\mathcal{Q}_n(t) = \tilde{\mathcal{Q}}_n(1 - \frac{2}{\overline{\lambda} + \underline{\lambda}}t)$. Then $\mathcal{Q}_n$ is a polynomial of degree n with $\mathcal{Q}_n(0) = 1$ and

$$\mathcal{Q}_n(\mathcal{A}) = \tilde{\mathcal{Q}}_n(\mathcal{M}).$$

Hence

(a.13) $$[[\mathcal{Q}_n(\mathcal{A})]] = [[\tilde{\mathcal{Q}}_n(M)]].$$

Now for such a polynomial $\tilde{Q}_n$

(a.14) $$[[\tilde{Q}_n(\mathcal{M})]] = \max_{\lambda \in \sigma(\mathcal{M})} |\tilde{Q}_n(\lambda)| = \max_{\lambda \in [-\rho, \rho]} |\tilde{Q}_n(\lambda)|,$$

where $\rho = \frac{\bar{\lambda} - \underline{\lambda}}{\bar{\lambda} + \underline{\lambda}} < 1$. From (a.12), (a.13) and (a.14)

(a.15) $$(Ae_n, e_n)^{1/2} \leq \max_{\lambda \in [-\rho, \rho]} |\tilde{Q}_n(\lambda)| (A(x - y_0), x - y_0)^{1/2}$$

for any $\tilde{Q}_n$ with $\tilde{Q}_n(1) = 1$. The best choice, with only the knowledge of ρ, is known to be given in terms of the Chebychev polynomial C_n; i.e.,

$$\tilde{Q}_n(\lambda) = C_n(\lambda/\rho)/C_n(1/\rho).$$

Note that this choice gives an upper bound without using the optimality property of C_n. Now C_n is defined as follows:

$$C_n(t) = \begin{cases} \cos(n \cos^{-1} t), & \text{if } |t| \leq 1 \\ \cosh(n \cosh^{-1} t), & \text{if } |t| > 1. \end{cases}$$

Hence

(a.16) $$\max_{\lambda \in [-\rho, \rho]} |\tilde{Q}_n(\lambda)| \leq 1/C_n(1/\rho).$$

Since $\rho < 1$, setting $\sigma = \cosh^{-1}(1/\rho)$, we have

$$C_n(1/\rho) = \frac{e^{n\sigma} + e^{-n\sigma}}{2} = e^{n\sigma} \left[\frac{1 + e^{-2n\sigma}}{2} \right].$$

Now

(a.17) $$1/C_n(1/\rho) = e^{-n\sigma} \left[\frac{2}{1 + e^{-2n\sigma}} \right] \leq 2e^{-n\sigma}.$$

But $\sigma = \ln[1/\rho + \sqrt{1/\rho^2 - 1}]$ so that

(a.18) $$e^{-n\sigma} = [1/\rho + \sqrt{1/\rho^2 - 1}]^{-n} = \left(\frac{\rho}{1 + \sqrt{1 - \rho^2}} \right)^n.$$

134

Now $K(A) = \bar{\lambda}/\underline{\lambda}$. Hence

$$\rho = \frac{K(A) - 1}{K(A) + 1}.$$

Combining (a.17) and (a.18), we find that

(a.19)
$$1/C_n(1/\rho) \le 2\left(\frac{K^{1/2}(A) - 1}{K^{1/2}(A) + 1}\right)^n.$$

Thus we have the desired estimate

$$(Ae_n, e_n)^{1/2} \le 2\left(\frac{K^{1/2}(BA) - 1}{K^{1/2}(BA) + 1}\right)^n (A(x - y_0), x - y_0)^{1/2}.$$

Note that, asymptotically, this rate is better than that of the simple linear iteration (1.3) since

$$\frac{K^{1/2}(BA) - 1}{K^{1/2}(BA) + 1} = \frac{K(BA) - 1}{K(BA) + 1 + 2K^{1/2}(BA)} \le \frac{K(BA) - 1}{K(BA) + 1}.$$

Derivation of the computational algorithm

We start by choosing y_0 and setting $r_0 = b - Ay_0$. Suppose now that y_n has been computed. We want to compute y_{n+1}. Set $r_n = b - Ay_n$.

a) If $r_n = 0$ then $y_n = x$ or $\hat{y}_n = x - y_0$.

Because $\tilde{V}_n \subset \tilde{V}_{n+1}$, $\hat{y}_j = x - y_0$ for $j \ge n$.

Hence $y_{n+1} = x$ and the process stops. Now define

$$\tilde{V}_n^\perp = \{y \in \tilde{V}_{n+1} \mid [Ay, z] = (Ay, z) = 0, \text{ for all } z \in \tilde{V}_n\}.$$

b) If $r_n \ne 0$ then $Br_n \ne 0$ and

$$Br_n = Br_0 - A\hat{y}_n = z_0 - A\hat{y}_n \in \tilde{V}_{n+1}.$$

But $(r_n, z) = (b - Ay_n, z) = 0$ for all $z \in \tilde{V}_n$ by (a.6). Hence $Br_n \notin \tilde{V}_n$. Therefore $\tilde{V}_n^\perp \ne \{0\}$. Now for any $p_n \in \tilde{V}_n^\perp$ with $p_n \ne 0$ we assert that

(a.20)
$$y_{n+1} = y_n + \alpha_n p_n$$

135

with

(a.21)
$$\alpha_n = (r_n, p_n)/(Ap_n, p_n).$$

Because of uniqueness we only need to check that (a.6) is satisfied. Since $\tilde{V}_{n+1} = \tilde{V}_n \oplus \tilde{V}_n^{\perp}$ and $\tilde{V}_n^{\perp}$ is one dimensional, any element $z_{n+1} \in \tilde{V}_{n+1}$ may be written as

$$z_{n+1} = z_n + \beta p_n$$

with $z_n \in \tilde{V}_n$. Thus

$$(A(y_n + \alpha_n p_n) - b, z_n + \beta p_n)$$
$$= (Ay_n - b, z_n) - \beta[(b - Ay_n, p_n) - \alpha_n(Ap_n, p_n)] + \alpha_n(Ap_n, z_n)$$
$$= 0,$$

since each term is zero. Hence (a.20) is proved.

Hence if we can find some $p_n \in \tilde{V}_n^{\perp}$ with $p_n \neq 0$ then we can compute y_{n+1} from y_n and p_n, using (a.20) and (a.21). Taking $p_0 = Br_0$, $p_0, \ldots, p_{n-1}$ are orthogonal with respect to $(A\cdot, \cdot)$ by construction and hence form a basis for $\tilde{V}_n$. Since Br_n has a component in $\tilde{V}_n^{\perp}$, we may take p_n to be the A-orthogonal projection of Br_n onto $\tilde{V}_n^{\perp}$; i.e.,

(a.22)
$$p_n = Br_n - \sum_{j=0}^{n-1} \frac{(Ap_j, Br_n)}{(Ap_j, p_j)} p_j.$$

But if $j \leq n - 2$ then $BAp_j = Ap_j \in \tilde{V}_n$. Hence $(Ap_j, Br_n) = (Ap_j, r_n) = 0$. Thus (a.22) reduces to

$$p_n = Br_n - \frac{(Ap_{n-1}, Br_n)}{(Ap_{n-1}, p_{n-1})} p_{n-1}.$$

The updated p_n is therefore given by a two term recurrence. We summarize the algorithm as follows:

Let y_0 be arbitrary. Set

$$r_0 = b - Ay_0$$

136

and

$$z_0 = Br_0 = p_0.$$

With y_n, r_n, z_n and p_n known, compute as follows for $n = 0, 1, \ldots$:

1) Ap_n

2) $y_{n+1} = y_n + \alpha_n p_n$, $\alpha_n = (r_n, p_n)/(Ap_n, p_n)$

3) $r_{n+1} = r_n - \alpha_n Ap_n$, (If $r_{n+1} = 0$, stop.)

4) $z_{n+1} = Br_{n+1}$

5) $p_{n+1} = z_{n+1} - \beta_n p_n$, $\beta_n = (Ap_n, z_{n+1})/(Ap_n, p_n)$.

This shows that the evaluations of A and B enter only once per step. If $B = I$ then this is just the usual CG algorithm.

It is easy to show that α_n and β_n may be computed alternatively as

$$\alpha_n = (r_n, z_n)/(Ap_n, p_n)$$

and

$$\beta_n = -(r_{n+1}, z_{n+1})/(r_n, z_n).$$

This way of computing saves one inner product computation and storage space for one vector.

The above discussion has been in terms of SPD operators A and B on M. We shall show how this abstract algorithm translates into a concrete matrix algorithm assuming that A and B are given in a certain way and that a concrete computational basis is used for representing the elements of the space M.

We shall consider the operator $A : M \to M$ as given by

$$(Au, \varphi) = A(u, \varphi),$$

for all $\varphi \in M$. The problem $Au = b$ corresponds to

$$(Au, \varphi) = (b, \varphi),$$

for all $\varphi \in M$. If $\{\varphi^i\}$ is a computational basis for M, then we seek $u = \sum_{j=1}^{N} u_j \varphi^j$ satisfying

$$\tilde{A}\tilde{u} = \tilde{F},$$

where $\tilde{u}$ is the vector $\{u_j\}$, $\tilde{F}$ is the vector $\{(b, \varphi^i)\}$ and $\tilde{A}$ is the matrix $\{A(\varphi^i, \varphi^j)\}$. Let $< \cdot, \cdot >$ be the Euclidean inner product. Then the relation between A and $\tilde{A}$ is

$$(Au, v) =< \tilde{A}\tilde{u}, \tilde{v} >,$$

where $\tilde{v} = \{v_j\}$ is the vector representing v in the basis $\{\varphi^j\}$.

We shall assume that we are given another SPD matrix $\tilde{B}$ so that $\tilde{B}\tilde{A}\tilde{u}$ corresponds to an SPD operator $\widetilde{BA}u = \tilde{B}\tilde{A}\tilde{u}$. As shown in Section 10, this is indeed the situation in the construction of $B = B_J$ in the multigrid algorithms.

In terms of $\tilde{A}$, $\tilde{B}$ and $< \cdot, \cdot >$ the preconditioned conjugate gradient algorithm is as follows. Let $\tilde{F} = \{(b, \varphi^i)\}$ be a given vector. Let $\tilde{y}_0$ be arbitrary and set

$$\tilde{r}_0 = \tilde{F} - \tilde{A}\tilde{y}_0$$

and

$$\tilde{z}_0 = \tilde{B}\tilde{r}_0 = \tilde{p}_0.$$

With $\tilde{y}_n$, $\tilde{r}_n$, $\tilde{z}_n$ and $\tilde{p}_n$ known, compute as follows for $n = 0, 1, \ldots$:

1) $\tilde{A}\tilde{p}_n$

2) $\tilde{y}_{n+1} = \tilde{y}_n + \alpha_n \tilde{p}_n$, $\alpha_n =< \tilde{r}_n, \tilde{z}_n > / < \tilde{A}\tilde{p}_n, \tilde{p}_n >$

3) $\tilde{r}_{n+1} = \tilde{r}_n - \alpha_n \tilde{A}\tilde{p}_n$, (If $\tilde{r}_{n+1} = 0$, stop.)

4) $\tilde{z}_{n+1} = \tilde{B}\tilde{r}_{n+1}$

5) $\tilde{p}_{n+1} = \tilde{z}_{n+1} - \beta_n \tilde{p}_n$, $\beta_n = - < \tilde{r}_{n+1}, \tilde{z}_{n+1} > / < \tilde{r}_n, \tilde{z}_n >$.

The preconditioned conjugate gradient algorithm is very often presented in terms of matrices by introducing the square root of the matrix $\tilde{B}$ and then showing that it can be implemented without computing the square root. The matrix algorithm here is mathematically identical to that algorithm.

Bibliographical Notes

The conjugate gradient method was first derived independently by Hestenes and Stiefel. This work was jointly published in [101]. Their work included the use of a preconditioner. The first use of preconditioning, using a linear iterative algorithm for partial differential equations (the alternating direction method), seems to have been given by Wachspress [158]. For a history and extensive annotated bibliography see the survey article by Golub and O'Leary [93].

Appendix B: Introduction to interpolation spaces

In this appendix we give a brief introduction to the theory of interpolation spaces using the so–called real method of interpolation of Lions and Peetre [110] [111]. We will restrict our attention to certain special cases which are the most relevant to the theory of these notes. We present some basic definitions and some pertinent consequences.

Let B_0 and B_1 be two Banach spaces with B_1 continuously embedded and dense in B_0. An <u>intermediate space</u> B is any subspace of B_0 satisfying

$$B_1 \subset B \subset B_0.$$

Real Method of Interpolation.

We want to define for $0 < s < 1$ a scale of spaces B_s with

$$B_1 \subset B_s \subset B_0,$$

and having nice properties. Define for each $t > 0$ and $u \in B_1$

(b.1)
$$K(t, u) = \inf_{u_0 + u_1 = u} (\|u_0\|_{B_0}^2 + t^2 \|u_1\|_{B_1}^2)^{1/2}$$

where $u_0 \in B_0$ and $u_1 \in B_1$. Note that we could define $K(t, u)$ by

$$\inf_{u_0 + u_1 = u} (\|u_0\|_{B_0} + t\|u_1\|_{B_1}).$$

The definition in (b.1) is necessary in the Hilbert space case in order that the intermediate spaces be Hilbert spaces.

For $0 < s < 1$ and $1 \le p < \infty$ define the quantity

$$\|\|u\|\|_{B_{s,p}} = \left(\int_0^\infty t^{-sp} K^p(t, u) \frac{dt}{t} \right)^{1/p}.$$

The case $p = \infty$ is defined by

$$\|\|u\|\|_{B_{s,\infty}} = \sup_{t>0} t^{-s} K(t, u).$$

It follows easily that $|||u|||_{B_{s,p}}$ is a norm on B_1. Now define

$$B_{s,p} = \{u \in B_0; |||u|||_{B_{s,p}} < \infty\}.$$

Then $B_{s,p}$ is a Banach space and is intermediate. The following is easily proved.

THEOREM B.1. For $u \in B_1$

(b.2) $$|||u|||_{B_{s,p}} \leq C_{s,p}\|u\|_{B_0}^{1-s}\|u\|_{B_1}^{s}$$

with $C_{s,p} = (ps(1-s))^{-1/p}$ if $p < \infty$ and $C_{s,\infty} = 1$.

Proof: For $u \in B_1$

$$K(t,u) \leq \|u\|_{B_0}$$

and

$$K(t,u) \leq t\|u\|_{B_1}.$$

Hence, for $p < \infty$ and $\alpha > 0$,

$$|||u|||_{B_{s,p}}^{p} \leq \left[\left(\int_0^\alpha t^{-1+p(1-s)}dt\right)\|u\|_{B_1}^{p} + \left(\int_\alpha^\infty t^{-1-ps}dt\right)\|u\|_{B_0}^{p}\right]$$

$$= \left[\frac{1}{p(1-s)}\alpha^{p(1-s)}\|u\|_{B_1}^{p} + \frac{1}{ps}\alpha^{-ps}\|u\|_{B_0}^{p}\right].$$

Now choose $\alpha = \frac{\|u\|_{B_0}}{\|u\|_{B_1}}$. Then

$$|||u|||_{B_{s,p}}^{p} \leq \frac{1}{ps(1-s)}\left(\|u\|_{B_0}^{1-s}\|u\|_{B_1}^{s}\right)^{p}$$

or

$$|||u|||_{B_{s,p}} \leq \|u\|_{B_0}^{1-s}\|u\|_{B_1}^{s}\left(\frac{1}{ps(1-s)}\right)^{1/p}.$$

For $p = \infty$

$$K(t,u) \leq \|u\|_{B_0}^{1-s}t^{s}\|u\|_{B_1}^{s}.$$

This proves Theorem B.1.

The Hilbert space case.

We shall consider the following situation. Suppose that $B_i = H_i$, $i = 0, 1$ are separable Hilbert spaces with $(\cdot, \cdot)$ the inner product on H_0. It is known that H_1 is the domain of an (unbounded) positive selfadjoint operator $\Lambda : H_1 \rightarrow H_0$ connecting the norms as follows:

$$\|u\|_{H_1} = \|\Lambda u\|_{H_0}.$$

For simplicity we shall consider the case where the spectrum of Λ is discrete and the eigenvectors form a complete orthonormal basis for H_0. Then we may expand any element of H_0 as

$$u = \sum_{i=1}^{\infty} (u, \varphi_i) \varphi_i.$$

If $u \in H_1$, then

$$\Lambda u = \sum_{i=1}^{\infty} \lambda_i (u, \varphi_i) \varphi_i$$

and

$$\|u\|_{H_1}^2 = \sum_{i=1}^{\infty} \lambda_i^2 (u, \varphi_i)^2.$$

We define the intermediate spaces H_s to consist of those elements of H_0 for which the norm

(b.3)
$$\|u\|_{H_s} = \left(\sum_{i=1}^{\infty} \lambda_i^{2s} (u, \varphi_i)^2 \right)^{1/2}$$

is finite. These spaces are clearly intermediate Hilbert spaces.

Now we want to connect these norms with the norms defined in terms of the real method of interpolation. We define

$$[[u]]_{H_s} = C_s \left(\int_0^{\infty} t^{-2s} K^2(t, u) \frac{dt}{t} \right)^{1/2}$$

for $0 < s < 1$ and $C_s = \left(\int_0^{\infty} \frac{t^{1-2s} dt}{t^2+1} \right)^{-1/2} = \sqrt{\frac{2}{\pi} \sin \pi s}$. The following representation result is quite important for our applications.

142

THEOREM B.2. *For $u \in H_s$, $0 < s < 1$,*

$$[[u]]_{H_s} = \|u\|_{H_s}.$$

Proof: Now we may write

$$K^2(t, u) = \inf_{u_1 \in H_1} (\|u - u_1\|_{H_0}^2 + t^2 \|u_1\|_{H_1}^2).$$

We solve the minimization problem. Let

$$u = \sum_{i=1}^{\infty} a_i \varphi_i$$

and

$$u_1 = \sum_{i=1}^{\infty} b_i \varphi_i.$$

Then

$$\|u - u_1\|_{H_0}^2 + t^2 \|u_1\|_{H_1}^2 = \sum_{i=1}^{\infty} [(a_i - b_i)^2 + t^2 \lambda_i^2 b_i^2].$$

We choose b_i to minimize $[(a_i - b_i)^2 + t^2 \lambda_i^2 b_i^2]$. The choice is $b_i = a_i (t^2 \lambda_i^2 + 1)^{-1}$.
Hence

$$K^2(t, u) = \sum_{i=1}^{\infty} t^2 \lambda_i^2 (t^2 \lambda_i^2 + 1)^{-1} a_i^2.$$

Now

$$\int_0^{\infty} t^{-2s} K^2(t, u) \frac{dt}{t} = \sum_{i=1}^{\infty} \left(\int_0^{\infty} t^{1-2s} \lambda_i^2 (t^2 \lambda_i^2 + 1)^{-1} dt \right) a_i^2$$

$$= C_s^{-2} \sum_{i=1}^{\infty} \lambda_i^{2s} a_i^2 = C_s^{-2} \|u\|_{H_s}^2.$$

Hence $[[u]]_{H_s} = \|u\|_{H_s}$, $0 < s < 1$.

The next result is the important property of "interpolation of operators". Suppose that we have spaces $\tilde{B}_0$ and $\tilde{B}_1$ analogous to B_0 and B_1; i.e., $\tilde{B}_1$ is continuously imbedded and dense in $\tilde{B}_0$. Define $\tilde{K}(t, \cdot)$ analogously. Further let L be a linear operator with $L : B_i \to \tilde{B}_i$ and constants C_i such that

(b.4) $$\|Lu\|_{\tilde{B}_i} \leq C_i \|u\|_{B_i}$$

for $u \in B_i$, $i = 0, 1$. Here $\| \cdot \|_B$ is the norm on the generic Banach space B.

THEOREM B.3. *Suppose that B_i and $\tilde{B}_i$ are as above and $L : B_i \to \tilde{B}_i$ satisfies (b.4). Then*

$$|||Lu|||_{\tilde{B}_{s,p}} \leq C_0^{1-s} C_1^s |||u|||_{B_{s,p}}.$$

Proof: Consider first $p < \infty$ and $0 < s < 1$. Then

$$|||Lu|||_{\tilde{B}_{s,p}} = \left(\int_0^\infty t^{-ps} \tilde{K}^p(t, Lu) \frac{dt}{t} \right)^{1/p}$$

$$\leq \left(\int_0^\infty t^{-ps} \inf_{u_0 + u_1 = u} \left(\|Lu_0\|_{\tilde{B}_0}^2 + t^2 \|Lu_1\|_{\tilde{B}_1}^2 \right)^{p/2} \frac{dt}{t} \right)^{1/p}$$

since $Lu = Lu_0 + Lu_1$ is a decomposition of Lu with $Lu_0 \in \tilde{B}_0$ and $Lu_1 \in \tilde{B}_1$. Using (b.4), we have

$$|||Lu|||_{\tilde{B}_{s,p}} \leq \left(\int_0^\infty t^{-ps} \inf_{u_0 + u_1 = u} \left(C_0^2 \|u_0\|_{B_0}^2 + t^2 C_1^2 \|u_1\|_{B_1}^2 \right)^{p/2} \frac{dt}{t} \right)^{1/p}$$

$$= C_0^{1-s} C_1^s \left(\int_0^\infty \left(\frac{C_1 t}{C_0} \right)^{-ps} K^p \left(\frac{C_1 t}{C_0}, u \right) \frac{dt}{t} \right)^{1/p}$$

$$= C_0^{1-s} C_1^s |||u|||_{B_{s,p}}.$$

The case $p = \infty$ is similar. The particular case in which the spaces are Hilbert spaces and the corresponding scale of spaces is defined by the spectral representation is important. The following is a combination of Theorems B.2 and B.3.

THEOREM B.4. *Let H_s and $\tilde{H}_s$ be Hilbert spaces defined as in (b.3) for $0 \leq s \leq 1$. Suppose that L is a linear operator with $L : H_i \to \tilde{H}_i$ and constants C_0 and C_1 such that*

$$\|Lu\|_{\tilde{H}_i} \leq C_i \|u\|_{H_i}$$

for $u \in H_i$, $i = 0, 1$. Then $L : H_s \to \tilde{H}_s$ and

$$\|Lu\|_{\tilde{H}_s} \leq C_0^{1-s} C_1^s \|u\|_{H_s}.$$

for $0 \leq s \leq 1$.

Proof: From the previous definitions

$$[[u]]_{H_s} = C_s |||u|||_{B_{s,2}} \text{ if } B_{s,2} = H_s.$$

Hence Theorem B.4 follows from Theorems B.2 and B.3.

144

Simultaneous approximation in scales of Banach spaces.

Let B_0 and B_1 be as above and suppose that $\{S_\epsilon\}$ are subspaces of B_1 depending on a parameter $\epsilon > 0$. Then we have the following:

THEOREM B.5. *Suppose that for any $p \geq 1$ and some θ with $0 < \theta < 1$*

$$(b.5) \qquad \inf_{\chi \in S_\epsilon} |||u - \chi|||_{B_{\theta,p}} \leq \epsilon^{1-\theta} \|u\|_{B_1},$$

for all $u \in B_1$. Then there is a constant $C(\theta) > 0$ such that

$$\inf_{\chi \in S_\epsilon} (\|u - \chi\|_{B_0} + \epsilon^\theta |||u - \chi|||_{B_{\theta,p}}) \leq C(\theta)\epsilon\|u\|_{B_1},$$

for all $u \in B_1$.

Proof: For $u \in B_1$, set $E(u) = \inf_{\chi \in S_\epsilon}(\|u - \chi\|_{B_0} + \epsilon^\theta|||u - \chi|||_{B_{\theta,p}})$, and $\delta = \sup_{\|u\|_{B_1} = 1} E(u)$. We shall show that $\delta \leq C(\theta)\epsilon$ which is the desired result. To this end let $v \in B_1$, $v \neq 0$. Then

$$(b.6) \qquad E(u) \leq \|u - v\|_{B_0} + \epsilon^\theta|||u - v|||_{B_{\theta,p}} + E(v)$$

$$= \|u - v\|_{B_0} + \epsilon^\theta|||u - v|||_{B_{\theta,p}} + \frac{E(v)}{\|v\|_{B_1}}\|v\|_{B_1}$$

$$\leq \|u - v\|_{B_0} + \delta\|v\|_{B_1} + \epsilon^\theta|||u - v|||_{B_{\theta,p}}.$$

Now choose v so that

$$(b.7) \qquad \|u - v\|_{B_0} + \delta\|v\|_{B_1} \leq 2K(\delta, u).$$

For such a v we can prove that for all s

$$(b.8) \qquad K(s, u - v) \leq \sqrt{10}K(s, u),$$

from which it follows that

$$(b.9) \qquad |||u - v|||_{B_{\theta,p}} \leq \sqrt{10}|||u|||_{B_{\theta,p}}.$$

To prove (b.8) for $\delta \leq s$ we see that, using (b.7),

$$K(s, u - v) \leq \|u - v\|_{B_0} \leq 2K(\delta, u) \leq 2K(s, u),$$

since $K(t, u)$ is an increasing function of t. For $s \leq \delta$, and any $w \in B_1$

$$K^2(s, u - v) \leq \|u - w\|_{B_0}^2 + s^2\|v - w\|_{B_1}^2$$

$$\leq \|u - w\|_{B_0}^2 + 2s^2\|w\|_{B_1}^2 + 2s^2\|v\|_{B_1}^2$$

$$\leq 2(\|u - w\|_{B_0}^2 + s^2\|w\|_{B_1}^2 + 4s^2\delta^{-2}K^2(\delta, u)),$$

where we used again (b.7). Since $t^{-1}K(t, u)$ is a decreasing function of t, we have

$$K^2(s, u - v) \leq 2(\|u - w\|_{B_0}^2 + s^2\|w\|_{B_1}^2 + 4K^2(s, u)).$$

Taking the infimum over w it follows that for $s \leq \delta$

$$K(s, u - v) \leq \sqrt{10}K(s, u).$$

Hence (b.8) holds. Using (b.7) and (b.9) with (b.6), we have

(b.10)
$$E(u) \leq 2K(\delta, u) + \sqrt{10}\varepsilon^\theta |||u|||_{B_{\theta,p}}$$

$$\leq 2\delta^\theta |||u|||_{B_{\theta,\infty}} + \sqrt{10}\varepsilon^\theta |||u|||_{B_{\theta,p}}.$$

It is an exercise to prove that, for $1 \leq p < \infty$,

(b.11)
$$|||u|||_{B_{\theta,\infty}} \leq p^{1/p}|||u|||_{B_{\theta,p}}.$$

We have from (b.10) and (b.11)

$$E(u) \leq (2p^{1/p}\delta^\theta + \sqrt{10}\varepsilon^\theta)|||u|||_{B_{\theta,p}}.$$

But $E(u) = E(u - \chi)$ for any $\chi \in S_\varepsilon$ so that, using (b.5),

$$E(u) \leq \sqrt{10}(\delta^\theta + \varepsilon^\theta) \inf_{\chi \in S_\varepsilon} |||u - \chi|||_{B_{\theta,p}} \leq \sqrt{10}(\delta^\theta + \varepsilon^\theta)\varepsilon^{1-\theta}\|u\|_{B_1}.$$

Hence

$$\delta \leq \sqrt{10}(\delta^\theta + \varepsilon^\theta)\varepsilon^{1-\theta}$$

Thus, for appropriate $C(\theta)$,

$$\delta \leq C(\theta)\varepsilon.$$

146

This proves the theorem.

Bibliographical Notes

The real method of interpolation of Lions and Peetre was given in [111] . It is described in the book of Lions and Magenes where they show the equality of the interpolation norm and the spectrally defined norm on the intermediate spaces. Theorems of the types of B.1 to B.4 may be found in many places (cf. [60]).

The approximation result Theorem B.5 is contained in [48].

Appendix C: Glossary of Conditions

A.1 $A(u,u) \leq C_a \sum_{k=1}^{J} A(\tilde{T}_k u, u)$, $\tilde{T}_1 = P_1$, $\tilde{T}_k = \lambda_k^{-1} A_k P_k$, $k > 1$.

A.1a Same as A.1 with $\tilde{T}_k$ replaced by $\hat{T}_k = \lambda_k^{-1} \hat{A}_k \hat{P}_k$.

A.2 $A(\tilde{T}_k w, w) \leq (\tilde{C} e^{k-\ell})^2 A(w,w)$, $0 < \varepsilon < 1$, $w \in M_\ell$, $\ell \leq k$.

A.2a Same as A.2 with $\tilde{T}_k$ replaced by $\hat{T}_k = \lambda_k^{-1} \hat{A}_k \hat{P}_k$.

A.3 $a_0 \frac{\|u\|^2}{\lambda_k} \leq (R_k u, u) \leq a_1 \frac{\|u\|^2}{\lambda_k}$, $R_k = R_k^t$, $u \in M_k$, $k \geq 2$.

A.3a Same as A.3 for $u \in \hat{M}_k$ and $R_k : M_k \to \hat{M}_k$.

A.4 $\frac{\|u\|^2}{\lambda_k} \leq C_R(\overline{R}_k u, u)$, $u \in M_k$, $\overline{R}_k = (I - K_k^* K_k) A_k^{-1}$

A.4 (alternate) $A(K_k v, K_k v) \leq A(K_{k,\omega} v, v)$

A.4a Same as A.4 for $u \in \hat{M}_k$, $R_k = R_k \hat{Q}_k$ and $R_k : M_k \to \hat{M}_k$.

A.5 $A(T_k u, T_k u) \leq \theta A(T_k u, u)$, $0 < \theta < 2$

A.6 There exist linear operators $\overline{Q}_k : M \to M_k$ with $\overline{Q}_J = I$ such that

$$\|(\overline{Q}_k - \overline{Q}_{k-1})u\|^2 \leq C\lambda_k^{-1} A(u,u)$$

and

$$A(\overline{Q}_k u, \overline{Q}_k u) \leq C A(u,u).$$

A.7 For some $\alpha \in (0,1]$

$$(A_k^{1-\alpha} Q_k u, Q_k u) \leq C_\alpha (A^{1-\alpha} u, u)$$

148

and

$$(A^{1-\alpha}(I - P_k)u, (I - P_k)u) \le C_\alpha \lambda_k^{-\alpha} A(u, u).$$

A.7a Assume that there exist linear operators $\overline{\overline{Q}}_k : M \to M_k$, $k = 1, \ldots, J$ with $\overline{\overline{Q}}_J = I$ with $\hat{M}_k \supseteq \text{Range}(\overline{\overline{Q}}_k - \overline{\overline{Q}}_{k-1})$. Assume further that for some $\alpha \in (0, 1]$

$$(\overline{\overline{A}}^{1-\alpha}(I - \overline{\overline{P}}_k)v, (I - \overline{\overline{P}}_k)v) \le C_\alpha \lambda_k^{-\alpha} A(v, v), \text{ for all } v \in M \subset \overline{\overline{M}}$$

and

$$(\overline{\overline{A}}_k^{1-\alpha} \overline{\overline{Q}}_k v, \overline{\overline{Q}}_k v) \le C_\alpha (\overline{\overline{A}}^{1-\alpha} v, v), \text{ for all } v \in M,$$

where λ_k is the largest eigenvalue of $\overline{\overline{A}}_k$ and $\overline{\overline{P}}_k$ and $\overline{\overline{Q}}_k$ are defined analogously.

A.8

$$\|(I - \overline{\overline{Q}}_k)u\| \le C\|(I - \overline{\overline{Q}}_k)u\|,$$

for all $v \in M_k$

A.9 $A_k(I_k u, I_k u) \le A_{k-1}(u, u)$, $u \in M_{k-1}$.

A.10 For some $\alpha \in (0, 1]$

$$|A_k((I - I_k P_{k-1})u, u)| \le C_\alpha^2 \left(\frac{\|A_k u\|_k^2}{\lambda_k}\right)^\alpha (A_k(u, u))^{1-\alpha}.$$

A.11 $A_k(I_k u, I_k u) \le 2A_{k-1}(u, u)$, $u \in M_{k-1}$.

A.12 For β_0 and β_1 with $1 < \beta_0 \le \beta_1$

$$\beta_0 m(k) \le m(k - 1) \le \beta_1 m(k), \; k = 2, \ldots, J.$$

A.13 For some number $\beta > 0$

$$|A(u, v) - \hat{A}(u, v)| \le Ch^\beta \|u\|_1 \|v\|_1,$$

for all u and $v \in M$.

Bibliography

1. R.A. Adams, "Sobolev Spaces," Academic Press, Inc., New York, 1975.
2. D.N. Arnold and F. Brezzi, *Mixed and nonconforming finite element methods: implementation, postprocessing and error estimates*, RAIRO Math. Model. Num. Anal. 19 (1985), 7–32.
3. G.P. Astrakhantsev, *Method for fictitious domains for a second-order elliptic equation with natural boundary conditions*, U.S.S.R. Comp. Math. and Math. Phys. 18 (1978), 114–121.
4. J.P. Aubin, "Approximation of Elliptic Boundary-Value Problems," Wiley-Interscience, New York, 1972.
5. O. Axelsson, *A generalized conjugate gradient, least squares method*, Numer. Math. 51 (1987), 209–228.
6. O. Axelsson and P.S. Vassilevski, *Algebraic multilevel preconditioning methods, II*, SIAM J. Numer. Anal. 27 (1990), 1569–1589.
7. A.K. Aziz and I. Babuška, *Part I, survey lectures on the mathematical foundations of the finite element method*, in "The Mathematical Foundations of the Finite Element Method with Applications to Partial Differential Equations," A.K. Aziz, ed., Academic Press, New York, NY, 1972, pp. 1–362.
8. I. Babuška, *On the Schwarz algorithm in the theory of differential equations of mathematical physics*, Tchecosl. Math. J. 8 (1958), 328–342 (in Russian).
9. I. Babuška, J.E. Osborn and J. Pitkäranta, *Analysis of mixed methods using mesh dependent norms*, Math. Comp. 35, 1039–1062.
10. N.S. Bakhvalov, *On the convergence of a relaxation method with natural constraints on the elliptic operator*, USSR Comp. Math. and Math. Phys. 6 (1966), 101–135.
11. R. Bank, *A comparison of two multilevel iterative methods for nonsymmetric and indefinite elliptic finite element equations*, SIAM J. Numer. Anal. 18 (1981), 724–743.
12. R.E. Bank and T. Dupont, *An optimal order process for solving finite element equations*, Math. Comp. 36 (1981), 35–51.
13. R.E. Bank, T.F. Dupont, and H. Yserantant, *The hierarchical basis multigrid method*, Num. Math. 52 (1988), 427–458.
14. R.E. Bank and C.C. Douglas, *Sharp estimates for multigrid rates of convergence with general smoothing and acceleration*, SIAM J. Numer. Anal. 22 (1985), 617–633.
15. R. Bank, J. Mandel and S. McCormick, *Variational multigrid theory*, in "Multigrid Methods," S. McCormick, ed., SIAM, Philadelphia, PA, 1987, pp. 131–178.
16. C. Bennet and R. Sharpley, "Interpolation of Operators," Academic Press, Inc., New York, 1988.
17. P.E. Bjørstad and O.B. Widlund, *Solving elliptic problems on regions partitioned into substructures*, "Elliptic Problem Solvers II," G. Birkhoff and A. Schoenstadt, eds., Academic Press, New York, 1984, pp. 245–256.
18. P.E. Bjørstad and O.B. Widlund, *Iterative methods for the solution of elliptic problems on regions partitioned into substructures*, SIAM J. Numer. Anal. 23 (1986), 1097–1120.
19. D. Braess and W. Hackbusch, *A new convergence proof for the multigrid method including the V-cycle*, SIAM J. Numer. Anal. 20 (1983), 967–975.
20. J.H. Bramble, *A second order finite difference analogue of the first biharmonic boundary value problem*, Numer. Math. 9 (1966), 236–249.
21. J.H. Bramble, *The Lagrange multiplier method for Dirichlet's problem*, Math. Comp. 37 (1981), 1–12.

22. J.H. Bramble, R.E. Ewing, R.R. Parashkevov and J.E. Pasciak, *Domain decomposition methods for problems with uniform local refinement in two dimensions*, in "Fourth Inter. Symp. on Domain Decomposition Methods for Partial Differential Equations," R. Glowinski, Y. Kuznetsov, G. Meurant, J. Périaux and O.B. Widlund, eds., SIAM, Phil. PA, 1991, pp. 91–100.

23. J.H. Bramble, R.E. Ewing, R.R. Parashkevov and J.E. Pasciak, *Domain decomposition methods for problems with partial refinement*, SIAM J. Sci. Stat. Comput. 13 (1992), 397–410.

24. J.H. Bramble, R.E. Ewing, J.E. Pasciak and A.H. Schatz, *A preconditioning technique for the efficient solution of problems with local grid refinement*, Comp. Meth. Appl. Mech. Eng. 67 (1988), 149–159.

25. J.H. Bramble, C. I. Goldstein and J.E. Pasciak, *Analysis of V-Cycle Multigrid Algorithms for Forms Defined by Numerical Quadrature*, in "Proceeedings of the Third Copper Mountain Conference, Multigrid Methods," S. McCormick, ed., Marcel Dekker, New York, 1992.

26. J.H. Bramble and S.R. Hilbert, *Estimation of linear functionals on Sobolev spaces with application to Fourier transforms and spline interpolation*, SIAM J. Numer. Anal. 7 (1970), 113–124.

27. J.H. Bramble and S. R. Hilbert, *Bounds for a class linear functionals on with application to Hermite interpolation*, Numer. Math. 16 (1971), 362–369.

28. J.H. Bramble, Z. Leyk, and J.E. Pasciak, *Iterative schemes for nonsymmetric and indefinite elliptic boundary value problems*, Math. Comp. 60 (1993), 1–22.

29. J.H. Bramble, Z. Leyk, and J.E. Pasciak, *The analysis of multigrid algorithms for pseudo-differential operators of order minus one*, Math. Comp. 63 (1994), 461–478.

30. J.H. Bramble and J.E. Pasciak, *An efficient numerical procedure for the computation of steady state harmonic currents in flat plates*, IEEE Trans. Mag. Mag-19 (1983), 2409–2412.

31. J.H. Bramble and J.E. Pasciak, *Preconditioned iterative methods for nonselfadjoint or indefinite elliptic boundary value problems*, in "Unification of finite element methods," H. Kardestuncer, ed., Elsevier Science Publ. (North-Holland), New York, 1984, pp. 167 – 184.

32. J.H. Bramble and J.E. Pasciak, *New convergence estimates for multigrid algorithms*, Math. Comp. 49 (1987), 311–329.

33. J.H. Bramble and J.E. Pasciak, *A preconditioning technique for indefinite systems resulting from mixed approximations of elliptic problems*, Math. Comp. 50 (1988), 1–18.

34. J.H. Bramble and J.E. Pasciak, *The analysis of smoothers for multigrid algorithms*, Math. Comp. 58 (1992), 467–488.

35. J.H. Bramble and J.E. Pasciak, *Uniform convergence estimates for multigrid V–cycle algorithms with less than full elliptic regularity*, in "Contemporary Mathematics Series 157," A Quarteroni, J. Periaux, Y.A. Kuznetsov and O.B. Widlund, eds., Am. Math. Soc., Prov. RI, 1994, pp. 17-26.

36. J.H. Bramble and J.E. Pasciak, *New estimates for multigrid algorithms including the V–cycle*, Math. Comp. 60 (1993), 447–471.

37. J.H. Bramble, J.E. Pasciak and A.H. Schatz, *An iterative method for elliptic problems on regions partitioned into substructures*, Math. Comp. 46 (1986), 361–369.

38. J.H. Bramble, J.E. Pasciak and A.H. Schatz, *The construction of preconditioners for elliptic problems by substructuring, I*, Math. Comp. 47 (1986), 103–134.

39. J.H. Bramble, J.E. Pasciak and A.H. Schatz, *The construction of preconditioners for elliptic problems by substructuring, II*, Math. Comp. 49 (1987), 1–16.

40. J.H. Bramble, J.E. Pasciak and A.H. Schatz, *The construction of preconditioners for elliptic problems by substructuring, III*, Math. Comp. 51 (1988), 415–430.

41. J.H. Bramble, J.E. Pasciak and A.H. Schatz, *The construction of preconditioners for elliptic problems by substructuring, IV*, Math. Comp. 53 (1989), 1–24.

42. J.H. Bramble, J.E. Pasciak, J. Wang, and J. Xu, *Convergence estimates for product iterative methods with applications to domain decomposition*, Math. Comp. 57 (1991), 1–21.

43. J.H. Bramble, J.E. Pasciak, J. Wang, and J. Xu, *Convergence estimates for multigrid algorithms without regularity assumptions*, Math. Comp. 57 (1991), 23–45.

44. J.H. Bramble, J.E. Pasciak and J. Xu, *The analysis of multigrid algorithms for nonsymmetric and indefinite elliptic problems*, Math. Comp. 51 (1988), 389–414.

45. J.H. Bramble, J.E. Pasciak and J. Xu, *Parallel multilevel preconditioners*, Math. Comp. 55 (1990), 1–22.

46. J.H. Bramble, J.E. Pasciak and J. Xu, *The analysis of multigrid algorithms with nonnested spaces or noninherited quadratic forms*, Math. Comp. 56 (1991), 1–34.

47. J.H. Bramble, J.E. Pasciak and J. Xu, *A multilevel preconditioner for domain decomposition boundary systems*, in "Proceedings of the 10'th Inter. Conf. on Comput. Meth. in Appl. Sci. and Engr.," Nova Sciences, New York, 1992.

48. J.H. Bramble and L.R. Scott, *Simultaneous approximation in scales of Banach spaces*, Math. Comp. 32 (1978), 947–954.

49. J.H. Bramble and J. Xu, *A local post-processing technique for improving the accuracy in mixed finite element approximations*, SIAM J. Numer. Anal. 24 (1989), 1267–1275.

50. J.H. Bramble and J. Xu, *Some estimates for weighted L^2 projections*, Math. Comp. 56 (1991), 463–476.

51. J.H. Bramble and M. Zlámal, *Triangular Elements in the Finite Element Method*, Math. Comp. 24 (1970), 809–820.

52. A. Brandt, *Multi-level adaptive technique (MLAT) for fast numerical solution to boundary value problems*, in "Proceedings of the 3rd Inter. Conf. on Numerical Meth. in Fluid Mech.," Springer–Verlag, Berlin and New York, 1973, pp. 82–89.

53. A. Brandt, *Multi-level adaptive solutions to boundary value problems*, Math. Comp. 31 (1977), 333–390.

54. A. Brandt, *Algebraic multigrid theory: the symmetric case*, Appl. Math. Comp. 19 (1986), 23–56.

55. S.C. Brenner, *An optimal-order multigrid method for P1 nonconforming finite elements*, Math. Comp. 52 (1989), 1–16.

56. S.C. Brenner, *An optimal-order nonconforming multigrid method for the biharmonic equation*, SIAM J. Numer. Anal. 26 (1989), 1124–1138.

57. F. Brezzi, *On the existence, uniqueness and approximation of saddle-point problems arising from Lagrange multipliers*, R.A.I.R.O. (1974), 129–151.

58. F. Brezzi, J. Douglas and L. Marini, *Two families of mixed finite elements for second order elliptic problems*, Numer. Math. 47 (1985), 217–235.

59. F. Brezzi and P.-A. Raviart, *Mixed finite element methods for 4th order elliptic equations*, in "Topics in Numerical Analysis III," Academic Press.

60. P.L. Butzer and H. Berens, "Semi-Groups of Operators and Approximation," Springer-Verlag, New York, 1967.

61. B.L. Buzbee and F.W. Dorr, *The direct solution of the biharmonic equation on rectangular regions and the Poisson equation on irregular regions*, SIAM J. Numer. Anal. 11 (1974), 753–763.

62. B.L. Buzbee, F.W. Dorr, J.A. George, and G.H. Golub, *The direct solution of the discrete Poisson equation on irregular regions*, SIAM J. Numer. Anal. 8 (1971), 722–736.

63. X.-C. Cai and O. Widlund, *Domain decomposition algorithms for indefinite elliptic problems*, SIAM J. Sci. Stat. Comp. 13 (1992), 243–258.

64. R. Chandra, "Conjugate Gradient Methods for Partial Differential Equations," Yale Univ., Dept. of Comp. Sci, Rep. No. 129, 1978.

65. P.G. Ciarlet, "The Finite Element Method for Elliptic Problems," North-Holland, New York, 1978.

152

66. P.G. Ciarlet and P.-A. Raviart, *A mixed finite element method for the biharmonic equation*, in "Mathematical Aspects of Finite Elements in Partial Differential Equations," Academic Press, New York, pp. 125–145.

67. P. Concus, G.H. Golub and G. Meurant, *Block preconditioning for the conjugate gradient method*, SIAM J. Sci. Stat. Comput. 6 (1985), 220–252.

68. R. Courant and D. Hilbert, "Methods of Mathematical Physics," Interscience, N.Y., 1962.

69. M. Dauge, "Elliptic boundary value problems on corner domains : smoothness and asymptotics of solutions," Lecture Notes in Math. # 1341, Springer-Verlag, New York, 1988.

70. N. Decker, J. Mandel and S. Parter, *On the role of regularity in multigrid methods*, in "Multigrid Methods," Proceeedings of the Third Copper Mountain Conference S. McCormick, ed., Marcel Dekker, New York, 1988.

71. Q.V. Dihn, R. Glowinski and J. Périaux, *Solving elliptic problems by domain decomposition methods*, in "Elliptic Problem Solvers II," G. Birkhoff and A. Schoenstadt, eds., Academic Press, New York, 1984, pp. 395–426.

72. C.C. Douglas, *Multi-grid algorithms with applications to elliptic boundary-value problems*, SIAM J. Numer. Anal. 21 (1984), 236–254.

73. M. Dryja and O.B. Widlund, *An additive variant of the Schwarz alternating method for the case of many subregions*, Technical Report, Courant Institute of Mathematical Sciences 339 (1987).

74. M. Dryja and O.B. Widlund, *Some domain decomposition algorithms for elliptic problems*, in "Iterative Methods for Large Linear Systems," L. Hayes and D. Kincaid, eds., Academic Press, New York, NY, 1989.

75. M. Dryja and O.B. Widlund, *Additive Schwarz methods for elliptic finite element problems in three dimensions*, Technical Report, Courant Institute of Mathematical Sciences 570 (June 1991).

76. T. Dupont, R.P. Kendall and H.H. Rachford, *An approximate factorization procedure for solving self-adjoint elliptic difference equations*, SIAM J. Numer. Anal. 5 (1968), 559–573.

77. T. Dupont and L.R. Scott, *Polynomial approximation of functions in Sobolev spaces*, Math. Comp. 34 (1980), 441–463.

78. S.C. Eisenstat, H.C. Elman, M.H. Schultz, and A.H. Sherman, *The (new) Yale sparse matrix package*, in "Elliptic Problem Solvers II," G. Birkhoff and A. Schoenstadt, eds., Academic Press, New York, 1984, pp. 45–52.

79. H.C. Elman, *Iterative methods for large, sparse, nonsymmetric systems of linear equations*, Yale Univ. Dept. of Comp. Sci. Rep. 229, (1982).

80. R.E. Ewing, R.D. Lazarov, P. Lu, and P.S. Vassilevski, *Preconditioning indefinite systems arising from mixed finite element discretization of second-order elliptic problems*, in "Preconditioned Conjugate Gradient Methods," O. Axelsson and L. Kolotilina, eds., Lecture Notes in Mathematics # 1457, Springer-Verlag, Berlin.

81. R.E. Ewing, R.D. Lazarov, T.F. Russell, and P.S. Vassilevski, *Local refinement via domain decomposition techniques for mixed finite element methods with rectangular Raviart-Thomas elements*, in "Domain Decomposition Methods," T.F. Chan, R. Glowinski, J. Periaux, and O.B. Widlund, eds., SIAM, Philadelphi, Penn., 1989, pp. 98 – 114.

82. R.S. Falk, *An analysis of the finite element method using Lagrange multipliers for the stationary Stokes equations*, Math. Comp. 30 (1976), 241–269.

83. R.S. Falk and J.E. Osborn, *Error estimates for mixed methods*, R.A.I.R.O. Numerical Analysis 14 (1980), 249–277.

84. R.P. Fedorenko, *The speed of convergence of one iterative process*, USSR Comput. Math. and Math. Phys. (1964), 1092–1096.

85. A. George and J.W. Liu, "Computer solution of large sparse positive definite systems," Prentice-Hall, Inc., Englewood Cliffs, N.J., 1981.

86. V. Girault and P.–A. Raviart, "Finite Element Approximation of the Navier-Stokes Equations," Lecture Notes in Math. # 749, Springer-Verlag, New York, 1981.

87. R. Glowinski, W. Kinton and M.F. Wheeler, *Acceleration of domain decomposition algorithms for mixed finite elements by multi-level methods*, in "Third Inter. Symp. on Domain Decomposition Methods for Partial Differential Equations," T. Chan, R. Glowinski, J. Periaux and O.B. Widlund, eds., SIAM, Phil. PA, 1990, pp. 263–289.

88. R. Glowinski and M.F. Wheeler, *Domain decomposition and mixed finite element methods for elliptic problems*, in "First Inter. Symp. on Domain Decomposition Methods for Partial Differential Equations," R. Glowinski, G.H. Golub, G.A. Meurant, and J. Periaux, eds., SIAM, Phil. PA, 1988, pp. 144–172.

89. R. Glowinski and M.F. Wheeler, *Domain decomposition and mixed methods for elliptic problems*, in "Proceedings, 1'st Inter. Conf. on Domain Decomposition Methods," SIAM, Philadelphia, 1988, pp. 144–172.

90. C.I. Goldstein, *Analysis and application of multigrid preconditioners for singularly perturbed boundary value problems*, SIAM J. Num. Anal. 26 (1989), 1090–1123.

91. C.I. Goldstein, *Multigrid analysis of finite element methods with numerical integration*, Math. Comp. 56 (1991), 409–436.

92. G.H. Golub and D. Meyers, *The use of preconditioning over irregular regions*, "Proc. 6th. Internl. Conf. Comput. Meth. Sci. and Engng., Versailles, FR.," 1983.

93. G.H. Golub and D.P. O'Leary, *Some history of the the conjugate gradient and Lanczos algorithms: 1948-1976*, SIAM Review 31 (1989), 50–102.

94. G. Golub and C.F. Van Loan, "Matrix Computations," Second Ed ition, The Johns Hopkins University Press, Baltimore.

95. P. Grisvard, *Behavior of the solutions of an elliptic boundary value problem in a polygonal or polyhedral domain*, in "Numerical Solution of Partial Differential Equations, III," B. Hubbard, ed., Academic Press, New York, 1976, pp. 207–274.

96. P. Grisvard, "Elliptic Problems in Nonsmooth Domains," Pitman, Boston, 1985.

97. W.D. Gropp and D.E. Keyes, *A comparison on Domain decomposition techniques for elliptic partial differential equations and the parallel implementation*, SIAM J. Sci. Stat. Comput. 8 (1987), s166–s203.

98. W. Hackbusch, "Multi-Grid Methods and Applications," Springer-Verlag, New York, 1985.

99. M.R. Hanisch, "Multigrid preconditioning for mixed finite element methods," Thesis, Cornell University, 1991.

100. M.R. Hestenes, *The conjugate gradient method for solving linear systems*, Proc. Symp. Appl. Math. VI, Amer. Math. Soc. (1956), 83–102.

101. M.R. Hestenes and E.S. Stiefel, *Methods of conjugate gradients for solving linear systems*, J. Res. Nat. Bur. Standards 49 (1952), 409–436.

102. M.M. Hrabok and T.M. Hrudey, *A review and catalog of plate bending finite elements*, Computers & Structures 19 (1984), 479–495.

103. C. Johnson and J. Pitkäranta, *Analysis of some mixed finite element methods related to reduced integration*, Math. Comp. 38 (1982), 375–400.

104. R.B. Kellogg, in "Interpolation between subspaces of a Hilbert space," Univ. of Maryland,, Inst. Fluid Dynamics and App. Math., Tech. Note BN-719, 1971.

105. J.T. King, *On the construction of preconditioners by subspace decomposition*, J. Comp. Appl. Math. 29 (1990), 192–205.

106. M. Kočvara and J. Mandel, *A multigrid method for three-dimensional elasticity and algebraic convergence estimates*, Appl. Math. Comp. 23 (1987), 121–135.

107. V.A. Kondrat'ev, *Boundary problems for elliptic equations with conical or angular points*, Trans. Moscow Math. Soc. 16 (1967), 227–313.

108. S.G. Krein and Y.I. Petunin, *Scales of Banach spaces*, Russian Math. Surveys 21 (1966), 85–160.

109. P. Lascaux and P. Lesaint, *Some nonconforming finite elements for the plate bending problem*, RAIRO Rev. Franc. D'Autom. 9 (1975), 9–54.

110. J.L. Lions and E. Magenes, "Non-Homogeneous Boundary Value Problems and Applications," Springer-Verlag, New York, 1972.

111. J.L. Lions and J. Peetre, *Sur une classe d'espaces d'interpolation*, Institut des Hautes Etudes Scientifique, Publ. Math. 19 (1964), 5–68.

112. P.L. Lions, *On the Schwarz alternating method*, in "In the Proceedings of the First International Symposium on Domain Decomposition Methods for Partial Differential Equations," R. Glowinski, G.H. Golub, G.A. Meurant, and J. Periaux, eds., 1987.

113. J.F. Maitre and F. Musy, *Algebraic formalization of the multigrid method in the symmetric and positive definite case - a convergence estimation for the V-cycle*, in "Multigrid Methods for Integral and Differential Equations," D.J. Paddon and H. Holstien, eds., Claridon Press, Oxford, 1985.

114. J. Mandel, *Étude algébrique d'une méthode multigrille pour quelques probléms de frontiére libre*, C.R. Acad. Sci., Sér. I. Math. 298 (1984), 469–472, Paris.

115. J. Mandel, *Multigrid convergence for nonsymmetric, indefinite variational problems and one smoothing step*, in "Proc. Copper Mtn. Conf. Multigrid Methods," Applied Math. Comput., 1986, pp. 201–216.

116. J. Mandel, *Algebraic Study of Multigrid Methods for Symmetric, Definite Problems*, Preprint.

117. J. Mandel and S.F. McCormick, *Iterative solution of elliptic equations with refinement: The two-level case*, in "Domain Decomposition Methods," T.F. Chan, R. Glowinski, J. Periaux, and O.B. Widlund, eds., SIAM, Philadelphi, Penn., 1989, pp. 81–92.

118. J. Mandel, S.F. McCormick and J. Ruge, *An Algebraic Theory for Multigrid Methods for Variational Problems*, SIAM J. Numer. Anal. 25 (1988), 91–110.

119. T.A. Manteuffel and S.V. Parter, *Preconditioning and boundary conditions*, SIAM J. Numer. Anal. 27 (1990), 656–694.

120. T.P. Mathew, "Domain Decomposition and Iterative Refinement Methods for Mixed Finite Element Discretizations of Elliptic Problems," Thesis, New York University, 1989.

121. S.F. McCormick, *Multigrid Methods for Variational Problems: Further Results*, SIAM J. Numer. Anal. 21 (1984), 255–263.

122. S.F. McCormick, *Multigrid Methods for Variational Problems: General Theory for the V-Cycle*, SIAM J. Numer. Anal. 22 (1985), 634–643.

123. S.F. McCormick and J. Ruge, *Unigrid for multigrid simulation*, Math. Comp. 41 (1983), 43–62.

124. S.F. McCormick and J. Thomas, *The fast adaptive composite grid (FAC) method for elliptic equations*, Math. Comp. 46 (1986), 439–456.

125. J.A. Meyerink and H.A. van der Vorst, *Iterative methods for the solution of linear systems of which the coefficient matrix is a symmetric M-matrix*, Math. Comp. 31 (1977), 148–162.

126. F.A. Milner, *Mixed finite element methods for quasilinear second-order elliptic problems*, Math. Comp. 44 (1985), 303–320.

127. L.S.D. Morley, *The triangular equilibrium element in the solution of plate bending problems*, Aero. Quart. 19 (1968), 149–169.

128. J. Nečas, *Sur la coercivité des formes sesquilinéares elliptiques*, Rev. Roumaine de Meth. Pure et Appl. 9 (1964), 47–69.

129. J. Nečas, "Les Méthodes Directes en Théorie des Équations Elliptiques," Academia, Prague, 1967.

130. J.C. Nedelec, *Elements finis miztes incompressibles pour l'equation de Stokes dans R^3.*, Numer. Math. 39 (1982), 97–112.

131. J.C. Nedelec and J. Planchard, *Une méthod variationnelle d'éléments finis pour la résolution numérique d'un problème extérieur dans R^3*, R.A.I.R.O. 7 (1973), 105–129.

132. R.A. Nicolaides, *On multiple grid and related techniques for solving discrete elliptic systems*, J. Computational Phys. **19** (1975), 418–431.

133. R.A. Nicolaides, *On the l^2 convergence of an algorithm for solving finite element equations*, Math. Comp. **31** (1977), 892–906.

134. J. Nitsche, *Ein Kriterium für die Quasi-Optimalität des Ritzchen Verfahrens*, Num. Math. **11** (1968), 346–348.

135. P. Oswald, *On discrete norm estimates related to multilevel preconditioners in the finite element method*, in "Constructive Theory of Functions," K.G. Ivanov, P. Petrushev and B. Sendov, eds., Proc. Int. Conf. Varna, Bulg. Acad. Sci., Sofia, 1991, pp. 203–214.

136. J.E. Pasciak, *Domain decomposition preconditioners for elliptic problems in two and three dimensions; first approach*, in "Proceedings, First International Symposium on Domain Decomposition Methods for Partial Differential Equations," R. Glowinski, G.H. Golub, G. A. Meurant, and J. Périaux, eds., SIAM, Philadelphia, 1987, pp. 62–72.

137. J.E. Pasciak, *Domain decomposition preconditioners for elliptic problems in two and three dimensions*, in "Numerical Algorithms for Modern Parallel Computer Architectures, IMA Volumes Math. Appl.," M. Schultz, ed., Springer-Verlag, New York, 1988, pp. 163–172.

138. J.E. Pasciak, *Two domain decomposition methods for Stokes equations*, in "Proceedings, Second International Symposium on Domain Decomposition Methods for Partial Differential Equations," T. Chan, R. Glowinski, J. Périaux and O.B. Widlund, eds., SIAM, Philadelphia, 1988, pp. 419–430.

139. W.M. Patterson, 3rd, "Iterative Methods for the Solution of a Linear Operator Equation in Hilbert Space - A Survey," Lecture Notes in Mathematics # 394, Springer-Verlag, New York, 1974.

140. P. Peisker, *A multilevel algorithm for the biharmonic problem*, Numer. Math. **46** (1985), 623–634.

141. P. Peisker and D. Braess, *A conjugate gradient method and a multigrid algorithm for Morley's finite element approximation of the biharmonic equation*, Numer. Math. **50** (1987), 567–586.

142. T. von Petersdorff and E.P. Stephan, *On the convergence of the multigrid method for a hypersingular integral equation of the first kind*, Numer. Math. **57** (1990), 379–391.

143. R. Rannacher, *On nonconforming and mixed finite element methods for plate bending problems. The linear case*, RAIRO Anal. Numer. **13** (1979), 369–387.

144. R. Rannacher, *Nonconforming finite element methods for eigenvalue problems in linear plate theory*, Numer. Math. **33** (1979), 23–42.

145. P.–A. Raviart and J.M. Thomas, *A mixed finite element method for 2-nd order elliptic problems*, in "Mathematical Aspects of Finite Element Methods," I. Galligani and E. Magenes, eds., Springer-Verlag, New York, 1977, pp. 292–315.

146. Y. Saad and M.H. Schultz, *GMRES: A generalized minimal residual algorithm for solving nonsymmetric linear systems*, SIAM J. Sci. Stat. Comput. **7** (1986), 856 – 869.

147. A.H. Schatz, *An observation concerning Ritz-Galerkin methods with indefinite bilinear forms*, Math. Comp. **28** (1974), 959–962.

148. H.A. Schwarz, *Ueber einige Abbildungsaufgaben*, J. Reine Angew. Math **70** (1869), 105–120 [Ges. Math. Abh., 2, 65–83. Springer,1890.].

149. L.R. Scott and M. Vogelius, *Conforming finite element methods for incompressible and nearly incompressible continua*, Inst. for Phys. Sci. and Tech., Univ. of Maryland, Tech. Rep. BN-1018.

150. L.R. Scott and S. Zhang, *Higher-dimensional nonnested multigrid methods*, Math. Comp. **58** (1992), 457–466.

151. G. Strang, *Approximation in the finite element method*, Numer. Math. **19** (1972), 81–93.

152. P.N. Swarztrauber, *The methods of cyclic reduction, Fourier analysis and the FACR algorithm for the discrete solution of Poisson's equation on a rectangle*, SIAM Review **19** (1977), 490–501.

153. R. Temam, "Navier-Stokes Equations," North-Holland Publishing Co., New York, 1977.

154. R.S. Varga, "Matrix Iterative Analysis," Prentice-Hall, Englewood Cliffs, NJ, 1962.

155. P. Vassilevski, *Iterative methods for solving finite element equations based on multilevel splitting of the matrix*, Bulgarian Academy Sciences, Sofia Bulgaria (1987).

156. R. Verfürth, *A multilevel algorithm for mixed problems*, SIAM J. Numer. Anal. 21 (1984), 264–284.

157. R. Verfürth, *A posteriori error estimates for the Stokes equations II non-conforming discretizations*, Numer. Math. 60 (1991), 235–249.

158. E.L. Wachspress, *Extended application of alternating direction implicit iteration model problem theory*, SIAM J. Appl. Math. 11 (1963), 994–1016.

159. J. Wang, *Convergence analysis without regularity assumptions for multigrid algorithms based on SOR smoothing*, SIAM J. Numer. Anal. 29 (1992), 987–1001.

160. J. Westlake, "A Handbook of Numerical Matrix Inversion and Solution of Linear Equations," John Wiley, New York, 1968.

161. O.B. Widlund, *Optimal iterative refinement methods*, Technical Report, Courant Institute of Mathematical Sciences 391 (1988).

162. J. Xu, *A new class of iterative methods for nonsymmetric boundary value problems*, SIAM J. Numer. Anal. 29 (1992), 303–319.

163. J. Xu, *Convergence estimates for some multigrid algorithms*, in "Proc. 1989 Houston Domain Decomp. Methods Conf. (oral presentation)."

164. J. Xu, "Theory of Multilevel Methods," Penn. State, Dept. Math. Rep AM-48, 1989.

165. J. Xu and X.–C. Cai, *A preconditioned GMRES method for nonsymmetric and indefinite problems*, Math. Comp. 59 (1992), 311–319.

166. H. Yserentant, *On the multi-level splitting of finite element spaces*, Numerische Mathematik 49 (1986), 379–412.

167. H. Yserentant, *The convergence of multi-level methods for solving finite-element equations in the presence of singularities*, Math. Comp. 47 (1986), 399–409.

168. S. Zhang, "Multi-level Iterative Techniques," Thesis, Penn. State Univ., 1988.

169. X. Zhang, *Multilevel Schwarz methods*, Numer. Math. 63 (1992), 521–539.

Index

159

Sobolev space, 95
Sobolev spaces, 9
SPD, 1
Spectrum, 2
Stephan, 107
Stiefel, 139
Stiffness matrix, 93, 107

Two level algorithm, 17

V–cycle, 62, 127
V–cycle algorithm, 22, 99
VanLoan, 8
Varga, 8
Variable V–cycle, 56, 66, 70, 109, 127
Verfürth, 119
von Petersdorff, 107

W–cycle, 56, 64, 65, 119, 127, 129
Wachspress, 139
Wang, 57, 88
Weak solution, 9
Widlund, 78
Work, 126

Xu, 20, 56, 57, 70, 84, 88

Zhang, 56, 57

Printed and bound by CPI Group (UK) Ltd, Croydon, CR0 4YY

23/10/2024

01778230-0007